Internet of Things

Technology, Communications and Computing

Series Editors

Giancarlo Fortino, Rende (CS), Italy

Antonio Liotta, Edinburgh Napier University,
School of Computing, Edinburgh, UK

The series Internet of Things - Technologies, Communications and Computing publishes new developments and advances in the various areas of the different facets of the Internet of Things.

The intent is to cover technology (smart devices, wireless sensors, systems), communications (networks and protocols) and computing (theory, middleware and applications) of the Internet of Things, as embedded in the fields of engineering, computer science, life sciences, as well as the methodologies behind them. The series contains monographs, lecture notes and edited volumes in the Internet of Things research and development area, spanning the areas of wireless sensor networks, autonomic networking, network protocol, agent-based computing, artificial intelligence, self organizing systems, multi-sensor data fusion, smart objects, and hybrid intelligent systems.

** Indexing:*Internet of Things* is covered by Scopus and Ei-Compendex **

More information about this series at http://www.springer.com/series/11636

G. R. Kanagachidambaresan

Role of Single Board Computers (SBCs) in rapid IoT Prototyping

 Springer

G. R. Kanagachidambaresan
Associate Professor, Department of CSE
Vel Tech Rangarajan Dr Sagunthala R&D Institute
of Science and Technology, Avadi
Chennai, Tamil Nadu, India

ISSN 2199-1073   ISSN 2199-1081   (electronic)
Internet of Things
ISBN 978-3-030-72959-2   ISBN 978-3-030-72957-8   (eBook)
https://doi.org/10.1007/978-3-030-72957-8

This Springer imprint is published by the registered company Springer Nature Switzerland AG
The registered company address is: Gewerbestrasse 11, 6330 Cham, Switzerland

Dedicated to my students, scholars, and family members.

Preface

The rapid growth of smart city and Industry 4.0 revolution have made Internet of Things and Edge analytics an inexorable precept for all engineering domains. The need for sophisticated and ambient environments has resulted in an exponential growth in automation, artificial intelligence-based methodologies, and technologies. The monitoring in factories and industries is mainly done through economical SBCs and low-range data transmission units. Rapid prototyping is used in industry for immediate deployment and to provide ad hoc solutions. Energy-Efficient System-on-Chips (SoC) and advanced Scale Integration techniques have resulted in edge computing to take necessary fast decisions. Edge computing and fog computing are the major sources of data gathering for data analytics purposes. The entire computing environments have been replaced by dumb terminals and single-board computers (SBCs) making investors secure and profitable. This book proposes how to work with SBCs for Internet of Thing (IoT) rapid prototyping purposes. Internet of Things possibly provides better solutions for ad hoc problems and sophisticates the lifestyle of the individual. This book describes the programming and interfacing of widely used SBCs such as Raspberry pi, Arduino, Beagle Bone, and STM boards.

This is an authored book providing novel programs to solve new technological real-time problems. This book addresses programming SBCs, PCB design, Mechanical CAD design helping readers to incorporate their ideas to Proof of Concept. The book aims at providing programming, interfacing of sensors, PCB design, and Mechanical Cad design and create rapid prototyping. This book presents the methodologies of rapid prototyping with Kicad design, and Catia software to create ready-to-mount solutions. The book covers scripting-based and drag/drop-based programming for different problems and data gathering approach. The book also discusses IoT projects and Industry 4.0 problems and solutions.

Chennai, India   G. R. Kanagachidambaresan

Acknowledgments

My heartfelt thanks to Editor in Chief and Managing Editor, Mary James for considering my candidature and providing me with the opportunity to author this book. I sincerely thank the management of Vel Tech Rangarajan Dr. Sagunthala R&D Institute of Science and Technology, Avadi, for supporting me in all occasions. I would like to extend my thanks to Dean R&D, Dean TBI, Vel Tech Nidhi Prayas, Mr. Shaju, Mr. V. Balaji and Dr. Silambarasan for providing space and opportunity. My Special thanks go to Mrs. V. Mahima and Mr. K. Ananthajith for their support and motivation, which has made this book possible. Special thanks to Satwat, Shefali, Anjali, Sowmiya, Rishu, Manendra, Research-Scholar V. Sowmiya, Mr. R. Rameshkumar, and friends.

Contents

About the Author

G. R. Kanagachidambaresan received his B.E. degree in Electrical and Electronics Engineering from Anna University in 2010 and M.E. Pervasive Computing Technologies in Anna University in 2012. He has completed his Ph.D. (Wireless Body Area Networks, Healthcare) in Anna University, Chennai, in 2017. He is currently an associate professor, Department of CSE, Vel Tech Rangarajan Dr. Sagunthala R&D Institute of Science and Technology. His main research interest includes body sensor network and fault-tolerant wireless sensor network. He has published several reputed articles and undertaken several consultancy activities for leading MNC companies. He has also guest edited several special issue volumes and books in SPRINGER and serving as editorial review board member for peer-reviewed journals. He is presently working on several government-sponsored research projects like ISRO, DBT, and DST. He is the editor in chief for Next Generation Computer and Communication Engineering Series, Wiley. He is also the managing director for Eazythings Technology Private Limited.

Chapter 1
Introduction to Internet of Things and SBCs

1.1 Introduction

The Single Board Computers (SBCs) normally consist to have all the processing and communication options in a single PCB board mainly manufactured to serve the purpose of low power and portability [1]. The present widely used SBCs include arduino, Raspberry Pi, Beagle Bone, Nvidia SBCs, Azus Tinker board, etc. [2–7]. These boards are operated with Wifi and in some cases COM port-based communications. The SBCs are operated with the linux-based OS, flashed over the sd cards. The OS image of the boards are normally downloaded from their respective websites and flashed in the SD card with win32 disk imager or etcher software. The boards are powered with USB-based 5 volt cables with more than 2 Amps current requirement to make it fully operational. A sincere caution is necessary for powering the SBC using apt chargers with proper power constraints. These devices in some cases are enabled with ipv4 wired connectivity. The boards like Raspberry pi 3, 3b+, Tinker board, and NVIDIA boards are given with 100 Mbps Ethernet connectivity options [8–10]. The boards like Raspberry pi, NVIDIA, Beagle bone, and azus tinker boards are in need of analog to digital converters to read real-time data from sensors [11].

Figure 1.1 illustrates the real-time SBCs operating in 5 V power supply. These devices are very small in size and capable for portable operations (Table 1.1).

G. R. Kanagachidambaresan, *Role of Single Board Computers (SBCs) in rapid IoT Prototyping*, Internet of Things, https://doi.org/10.1007/978-3-030-72957-8_1

Table 1.1 Size description of the potential single board computer widely used

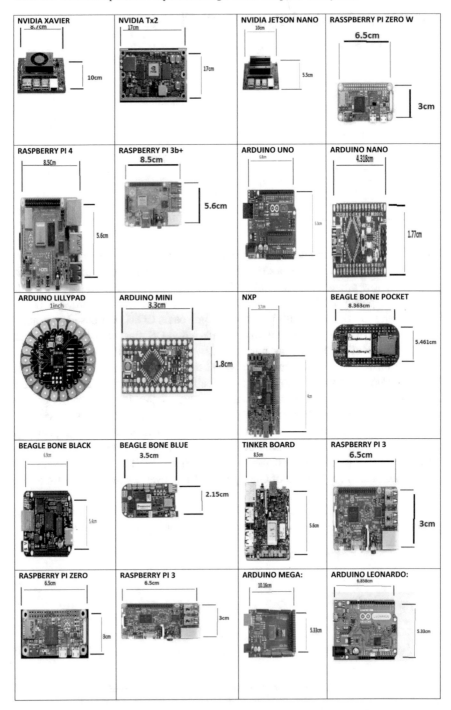

Table 1.2 provides the operational details of the boards, including OS, Power, Memory, system speed, and language.

Table 1.2 SBCs GPIOs and other necessary characteristics description

Board	Gpio	Os	Power	Memory	System speed	Basic language support
RASPBERRY PI 3	40 pin	Ubuntu Mate, Snappy Ubuntu Core, the Kodi-based media centers OSMC and LibreElec, the non-Linux-based Risc OS, It can also run Windows 10 IoT Core	5 V, 2 A min	1 GB RAM	RISC processor Speed 700 MHz to 1.4 GHz	C, C++, and Python
RAS PI 3b+	40 pin	Raspbian, UbuntuRetropieManjaro, OSMCLakka Kali Linux, Kano OS, GentooRecalBox, DietPi, LibreELEC Fedora, OpenMediaVault, CentOS	2.5 A	512 MB RAM	Processor speed 1.4 GHz	C,C++, and python
RAS PI 4	40-pin	Raspbian. Ubuntu MATE. Ubuntu Core. Ubuntu Server. Windows 10 IoT Core. Open Source Media Center.	3.0 A	2 GB, 4 GB, 8 GB RAM	Processor speed 1.5 GHz	C, C++, Python
RAS PI ZERO	40-pin	Raspbian Jessie, Kodi (aka OpenElec or OSMC) RetroPie. Kali Linux. MusicBox. MotionEyeOS. RuneAudio. ArchLinux.	1.2 A	512 MB RAM	Processor speed 1 GHz	C,C++, and python
RAS PI ZERO W	40 pin	Raspbian Jessie, Kodi (aka OpenElec or OSMC) RetroPie. Kali Linux. MusicBox. MotionEyeOS. RuneAudio. ArchLinux.	1.2 A	512 MB RAM	single-core ARM11 processor running at 1 GHz	C, C++, and Python
TINKER BOARD	40-pin	TinkerOS	5 V/3A	2 GB DDR3 RAM	up to 2.6 GHz turbo clock speed	C, C++, Python
NVIDIA JETSON NANO	40-pin	Linux OS	5 V/2A	4 GB RAM	1600 MHz	C, C++, Python
RASPBERRY pico	26 GPIO	Micropython,	5 V	Arm Cortex-M0+ processor, 264 kB, 16 MB of off-chip Flash	133 MHz	Python

Various SBCs are available; however mostly used economic SBC is RPi for prototyping purpose. The following section deals about accessing GPIOs of RPi. Similar concepts are applicable for other boards as well. Figure 1.1 provides the RPi GPIO Pin set starting from 3.3 V.

The General Purpose Input/Output pins in Rpi is classified as

- Digital Input/output
- Communication pins
- PWM outputs

The digital input pins are capable to read 0/1 states; however the raspberry pi port is not equipped with the analog and needs separate ic to read analog voltage.

There are about 40 pins in the raspberry pi recent versions. These pins are capable to provide 3.3 V output. Two 5 V pins are available; these pins are controlled with C, C++, Python, and node-red programming languages. The command to read the entire pin configuration in Pi is to type the following command in the pi terminal window.

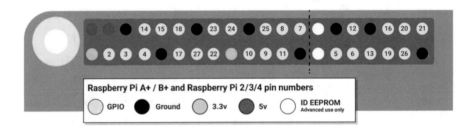

Fig. 1.1 GPIO pin numbering in RPi

```
gpioreadall
```

Figure 1.2 illustrates the output of the gpioreadall command. The average current that can be drawn from a gpio pin is about 16 mA.

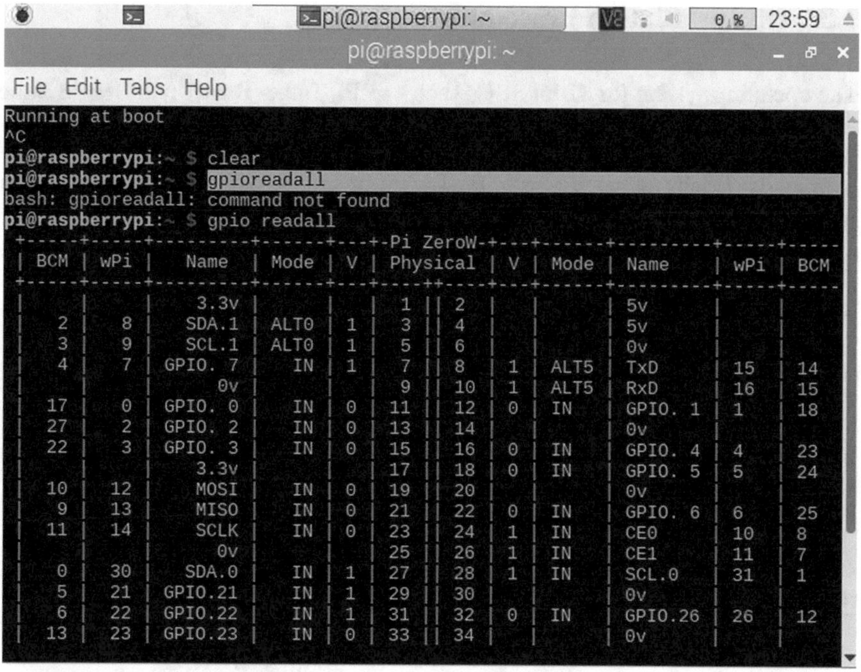

Fig. 1.2 Gpioreadall information in terminal window

There are certain dos and don'ts followed while using these gpio pins; some of these gpio pins are directly connected with the processor. Any wrong connection with these pins could entirely crash the OS and crash the hardware. In some of the SBCs, voltage protection circuit is not available; it is forbidden to connect source more than 3.3 V across any GPIO pins. Driving capacitive loads are not advisable with these pins; the system only should be connected with Resistor-Capacitor loads.

1.2 Hardwired Programming

Hardwire program is basically logic gate-based programming faster than micro programmed. It is very difficult, since all the realizations have to be done with the help of logic controls and control signals. The hardwired program is difficult to modify and possible to some limited extent.

It is advisable to use C programming for faster execution.

Code below illustrates the code snippet of the C programming used to trigger a led bulp in the RPi. The circuit is connected as per the code snippet connected.

1.3 OS Installation Procedure

The operating system for 32 bit SBCs such as RPi, Tinker Board, and Beagle Bone are collected from their respective website links and img file is flashed in the memory card, 32 Gb and above size, class 10 (for high speed) using win32 disk imager software or balena etcher software. Figure 1.3 illustrates the software os download links of raspberry pi, Tinker board, and Beagle bone board.

Fig. 1.3 OS installation procedure in Raspberry Pi

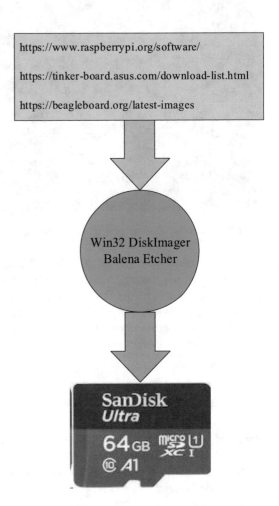

```
#include <wiringPi.h>
#include <stdio.h>
#define Pin 8
int main(void){
pinMode(Pin, OUTPUT);
while(1){
        digitalWrite(Pin, LOW); //led on
        printf("Led on");
        delay(1000);        //pause programming for one second
        digitalWrite(Pin, HIGH);//led off
        print("Led off");
        delay(1000);
        }
return 0;
}
```

Code 1 C Code for RPi led operation

A delay of 1000 ms (i.e.) 1 s is given, it can be increased in delay loop. The following command is used to execute the above c file to blink a led for a second.

This is executed inside a while loop to make it infinity times.

```
gccblink.c -o led -1wiringPi
./led
```

1.4 IDEs and Programming Environment

The commonly used Integrated Development Environment (IDEs) for SBCs are discussed here. The SBCs which are 32 bit can work with linux-based operating systems. In some cases the boards are hard metal programmed to server very particular applications. Lightweight operating system is also available for boards such as raspberry pi to reduce the booting time in the application layer. Table 1.3 illustrates the type of board with their IDEs.

Table 1.3 IDE commonly used for single board computer

S. No	Nature of Board	Operating IDE
1.	Arduino	Arduino Ide, MATLAB, Keil C
2.	Raspberry Pi	Python, C, scratch, Node-red, etc., C, C++, etc.
3.	NXP	mbed, keil c programming, C, C++, etc.
4.	Tinker	Python, node-red, scratch, C, C++, etc.
5.	NVIDIA	Python, Node-red, CUDA, Scratch, etc.

The following section discusses some of the widely used compilers.

1.5   Keil C Programming

Programming Language Keil originated from German—ARM holdings. The year of establishment of Keil is 1982, done by Gunter and Reinhard Keil. The organization was renamed on April 1985 to Keil Elektronic GmbH; this gave fabulous improvements for silicon industries. The Keil succeeded in main c compiler on 8051 microcontroller which attracted them a huge respect from Electronics Industries. The Keil broad spectrum tools include ANSI C, debugers, MACRO assemblers, processor simulators, IDEs, program linkers, RTOS, etc.

1.5.1   Keil Execution

Keil is given under an open-source permit; RTX CMSis-RTOS can be freely created and shared by the software components that require a real-time operating system. It is also free for software developers working with STM32 gadgets.

Nowadays, microcontrollers are programmed by the keil C and also called as embedded C. Figure 1.4 explains the flashing steps involved using Keil C, The compiler converts the C program to hex codes and programmer unit programs to the target microcontroller.

1.6   Embedded System Programing Diagram

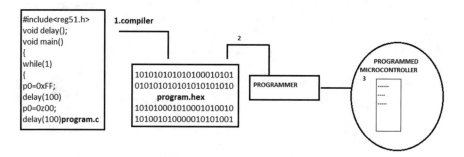

```
#include<reg51.h>        1.compiler
void delay();
void main()
{
while(1)
{                        101010101010100010101
p0=0xFF;                 010101010101010101010
delay(100)                      program.hex
p0=0z00;                 101010001010001010010
delay(100)program.c      101001010000010101001
```

PROGRAMMER

PROGRAMMED
MICROCONTROLLER
3

Fig. 1.4 Embedded system programming

Some of embedded c keywords:

- sbit
- bit
- SFR
- volatile
- macros define

1.7   Programming Arduino from MATLAB

The arduino boards can also be controlled with the MATLAB hardware support packages. The hardware can be communicated with com port communication. The simulink supports read sensor data and can operate actuators from digital pins. Figure 1.5 provides the hardware support package to be installed for communicating with the arduino controller.

Fig. 1.5 Arduino support package

Once the package is installed, the corresponding connected arduino can be invoked through a = arduino('COM5', 'uno') command. On successful exec, Fig. 1.6 appears on the command window. The led connected or default board led can be blinked with the following m-file in Matlab.

```
Command Window
New to MATLAB? See resources for Getting Started.
    Arduino Uno detected.
    This device is ready for use with MATLAB Support Package for Arduino Hardware. Get started with examples and other documentation.
    To use this device with Simulink, install Simulink Support Package for Arduino Hardware.

    >> a=arduino()
    Updating server code on board Uno (COM25). This may take a few minutes.

    a =

      arduino with properties:

                    Port: 'COM25'
                   Board: 'Uno'
           AvailablePins: {'D2-D13', 'A0-A5'}
    AvailableDigitalPins: {'D2-D13', 'A0-A5'}
       AvailablePWMPins: {'D3', 'D5-D6', 'D9-D11'}
    AvailableAnalogPins: {'A0-A5'}
     AvailableI2CBusIDs: [0]
               Libraries: {'I2C', 'SPI', 'Servo'}

fx >> |
```

Fig. 1.6 Arduino import in MATLAB command window

```
% create an arduino object
a = arduino();
for j=1:1:10
    writeDigitalPin(a, 'D13', 1);
    pause(1);
    writeDigitalPin(a, 'D13', 0);
    pause(1);
end
clear a
```

The default led pin is activated with 1 s delay for 10 times; once the loop completes, the arduino is disconnected.

1.8 SG90 Servo Control From MATLAB

Figure 1.7 provides the connection diagram for arduino and Matlab.

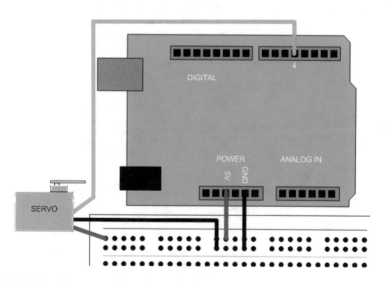

Fig. 1.7 Arduino Servo motor drive with MATLAB

The code to operate servo is as given below. Here the pulse width (off duty and on duty) is termed inside the servo function.

```
a = arduino('COM5', 'Uno', 'Libraries', 'Servo');
d = servo(a, 'D4', 'MinPulseDuration', 700*10^-6, 'MaxPulseDuration', 2300*10^-6);
for angle = 0:0.2:1
    writePosition(d, angle);
    current_pos = readPosition(d);
    current_pos = current_pos*180;
    fprintf('Current motor position is %d degrees\n', current_pos);
    pause(2);
end
```

Similarly, other sensors and drives can also be programmed by calling necessary library functions.

1.9   Mbed Programming

Figure 1.8 shows the STM32 Nucleo board with mbed programming support. The board can be connected with USB cable and programs can be flashed through online compiler.

Fig. 1.8 Mbed STM32 Nucleo board

mbed programming language provides open-source operating system for IoT with communication option and development options for more than 150 micro controller Unit nxp and some of the STM boards. Figure 1.9 shows the user palettes in the GUI of mbed online compiler.

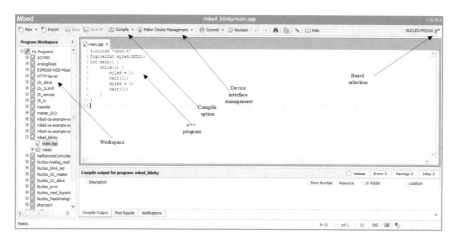

Fig. 1.9 Mbed UI design and components

Figure 1.10 shows the user interface to select the target board to flash the hex code. Once the program is written, it is compiled for error. The error-free program is ready to be flashed inside the mbed controller through the below window.

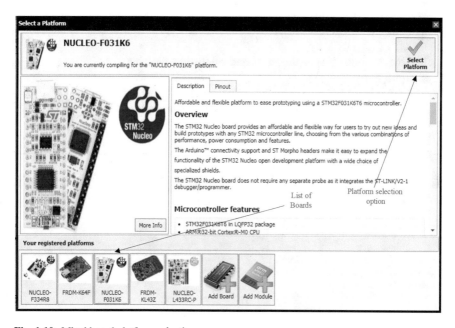

Fig. 1.10 Mbed board platform selection

The mbed program to flash a LED is given in the below code snippet.

```
#include "mbed.h"
DigitalOutmyled(LED2);
int main() {
while(1) {
myled = 1;
wait(1);
myled = 0;
wait(1);
    }
}
```

The mbed code to read analog values is given in the code snippet below.

1.9.1   Mbed Analog Read Programming

```
#include "mbed.h"
analogInpot(A0); //declares an Analog In
serial pc(USBTX, USBRX);
int main() {
while(1) {
float input = pot; //Reads the value and asigns to a variable
pc.printf("%c\n", input);
    //led = input; //Asign the value of the variable to the led
wait(0.05); //waits a 100 ms to set thae value on the led
   }
}
```

The mbed GPIOs are capable to generate PWM signals; the code given below provides the information to generate PWM signals.

1.9.2   Mbed PWM Generator Program

```
#include "mbed.h"
PwmOutmypwm(PWM_OUT);
DigitalOutmyled(LED1);
int main() {
mypwm.period_ms(10);
mypwm.pulsewidth_ms(1);
printf("pwm set to %.2f %%\n", mypwm.read() * 100);

while(1) {
myled = !myled;
wait(1);
  }
}
```

1.10   Scratch

1.10.1   Introduction

Scratch is a visual programming language that gives a wide range of provision for students to design their own interactive stories, games, and animations. Scratch was discovered by the Lifelong Kindergarten group at MIT Media Lab. As students design Scratch projects, they learn to think creatively, reason systematically, and work collaboratively. The scratch programming language is essentially focused on children aged eight and older and is intended to teach computational thinking using an easy but effective building block approach to software development that gives important on problem-solving than on specific syntax.

1.10.2   User Interface

Scratch is available for free download at http://scratch.mit.edu. Internet is required to download scratch, and once it is done, user does not need to bother about internet connection to create a project .This interface is split into three effective sections that are stage, area, block palette, and a coding area to layout the blocks into scripts and the green flag is used as a run flag another way to run the program is by clicking the code itself and the user are allowed to create their program block that will be appearing in the "my blocks."

The stage area resultant in animation, turtle graphics both in small and normal size with the full screen this stage uses x and y coordinates. The series of commands can be given in the coding area by dragging them from the block palette in the coding area. To create various effects of the sprite costumes tabs are used that includes animation and sound tabs. It allows user to add sound and music. Users are allowed to create sprites and background or draw on their own sprite manually. They can also choose from the library or upload an image; programming blocks are motion, looks, sound, event, control, sensing, operators, variable, my block, and extension.

Figure provides the scratch GUI in Raspberry pi. The raspberry pi OS comes with default scratch installation (Fig. 1.11).

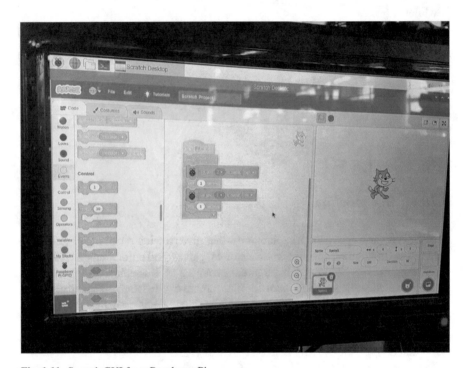

Fig. 1.11 Scratch GUI from Raspberry Pi

Figure 1.12 provides the drag and drop code on scratch.

Fig. 1.12 Scratch
programming

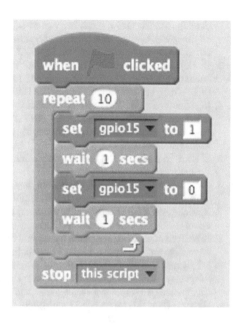

1.10.3 Offline Editing, Extensions, Code Base

Offline creation and playing of scratch are allowed, extension adds extra blocks, Scratch 3.0 is a new JavaScript-based codebase built of various component scratch GUI, scratch VM.

1.11 Conclusion

The widely used economic SBCs are discussed in this chapter. The SBCs in the present market do not need any additional programmers to program the boards. The size, programming abilities are discussed in detail in this chapter. Glimpses on different available IDEs are also discussed in this chapter.

References

1. P. Abrahamsson, et al., "Affordable and Energy-Efficient Cloud Computing Clusters: The Bolzano Raspberry Pi Cloud Cluster Experiment," in *2013 IEEE 5th International Conference on Cloud Computing Technology and Science*, Bristol, United Kingdom, Dec. 2013, pp. 170–175. https://doi.org/10.1109/CloudCom.2013.121

2. F. Salih, M.S.A. Omer, "Raspberry pi as a Video Server," in *2018 International Conference on Computer, Control, Electrical, and Electronics Engineering (ICCCEEE)*, Khartoum, (Aug. 2018), pp. 1–4. https://doi.org/10.1109/ICCCEEE.2018.8515817

3. O. Haffner, E. Kucera, M. Bachurikova, "Proposal of weld inspection system with single-board computer and Android smartphone," in *2016 Cybernetics & Informatics (K&I)*, Levoca, (Feb. 2016), pp. 1–5. https://doi.org/10.1109/CYBERI.2016.7438600

4. M. Radulovic, M. Pavlovic, N. Nenadic, G. Dimic, "Industrial Single Board Computer based on OMAP5 processor," in *2013 2nd Mediterranean Conference on Embedded Computing (MECO)*, Budva, Montenegro, (Jun. 2013), pp. 84–87. https://doi.org/10.1109/MECO.2013.6601324

5. S. Rani, R. Maheswar, G.R. Kanagachidambaresan, P. Jayarajan, *Integration of WSN and IoT for smart cities* (Springer, Cham, 2020)

6. *Programming with Tensorflow: solution for edge computing applications.* S.l.: Springer, 2020. https://link.springer.com/book/10.1007%2F978-3-030-57077-4

7. R. Maheswar, G. R. Kanagachidambaresan, R. Jayaparvathy, S. M. Thampi (eds.), *Body area network challenges and solutions* (Springer International Publishing, Cham, 2019)

8. P. Basford et al., Erica the rhino: a case study in using raspberry pi single board computers for interactive art. Electronics **5**(4), 35 (2016). https://doi.org/10.3390/electronics5030035

9. M. Dyvak, V. Tymets, "Emulation of programming environment for single-board computer Raspberry Pi at Monitoring the recurrent laryngeal nerve," in *2017 XIIIth International Conference on Perspective Technologies and Methods in MEMS Design (MEMSTECH)*, Lviv, Ukraine, (Apr. 2017), pp. 35–37. https://doi.org/10.1109/MEMSTECH.2017.7937527

10. Y. Li, J. Cheng, X. Wang, "An Optophone Based on Raspberry Pi and Android Wireless Communication," in *2020 IEEE International Conference on Advances in Electrical Engineering and Computer Applications(AEECA)*, Dalian, China, (Aug. 2020), pp. 952–956. https://doi.org/10.1109/AEECA49918.2020.9213587

11. S.J. Johnston et al., Commodity single board computer clusters and their applications. Future Gener Comput Syst, 12 (2018)

Chapter 2
Programming SBCs Using Python

2.1 Introduction

Most of the economic SBCs can be programmed with the Python language. This chapter provides the basic programming with necessary pip installation help.

String declaration, Arithmetic operation, decision-making, Looping concept, File operation, Introduction to matplotlib, RPi.GPIO python packages [1–6].

int(x) x is converted to plain integer.

long(x) x is converted to long integer.

float(x) x is converted to floating-point number.

complex(x)x is converted to complex number with real part x and imaginary part zero.

complex(x, y) x is converted to x and y to a complex number with real part x and imaginary part y. x and y are numeric expressions

Simple python programs are listed below for warming up the programming knowledge of the readers.

List programs

l1 = ['Maths1', 'Maths2', 17, 300];

l2 = [1, 2, 3, 4, 5];

l3 = ["a", "b", "c", "d"]

print(l1[0])

print(l2[1:5])

G. R. Kanagachidambaresan, *Role of Single Board Computers (SBCs) in rapid IoT Prototyping*, Internet of Things, https://doi.org/10.1007/978-3-030-72957-8_2

deleting value from list

```
l1 = ['Maths1', 'Maths2', 1997, 2000];
print l1
del l1[2];
print "After deleting value at index 2 : "
print l1
```

```
Hello world program
print("hello world")
Simple addition program
```

```
num1 = 3
num2 = 6
print(num1+num2)
```

Getting User Input (value)

```
num1 = int(input("enter number 1"))
num2 = float(input("enter number 2"))
print(num1+num2)
```

Getting User Input(String)

```
num1 = raw_input("enter number 1")
num2 = raw_input("enter number 2")
print(num1+num2)
#prints combined strings
```

2.2   File Operations

Write operation in file

```
fo1 = open("foo.txt", "wb")
print "Name of the file: ", fo1.name
fo1.write( "IoT is cool");
fo1.close()
```

Read operation in file

```
fo1 = open("foo.txt", "r+")
str = fo1.read(11);
print "Read String is : ", str
# Close opend file
fo1.close()
```

Exception handling in python

```
try:
fh = open("foo.txt", "r")
fh.write("This is my test file")
exceptIOError:
print "Error: can\'t find file"
else:
print "content written successfully"
```

Loops

while loop

```
i = 1
while (i <= 5):
print "i =", i
    i += 1
print "completed"
```

for loop

```
tot = 1
for i in range (1, 6, 1):
tot = tot + i
print "i = ", i, "\t tot = ", tot
print "completed"
```

if loop

```
weight = float(input("what is the weight "))
if weight > 100:
print("Heavy luggage")
print("Thank you ")
```

nested loop

```
for i in range (1, 3, 1):
for j in range (1, 4, 1):
print "i = ", i, " j = ", j
print "completed"
```

if else loop

```
temp = float(input('What is the outside temperature'))
if temp > 70:
print('cool yourself')
else:
print('wear warm clothes')
```

```
import pandas as pd
titanic_data = pd.read_csv('titanic.csv')
titanic_data.head()
```

	PassengerId	Survived	Pclass	Name	Sex	Age	SibSp	Parch	Ticket	Fare	Cabin	Embarked
0	1	0	3	Braund, Mr. Owen Harris	male	22.0	1	0	A/5 21171	7.2500	NaN	S
1	2	1	1	Cumings, Mrs. John Bradley (Florence Briggs Th...	female	38.0	1	0	PC 17599	71.2833	C85	C
2	3	1	3	Heikkinen, Miss. Laina	female	26.0	0	0	STON/O2. 3101282	7.9250	NaN	S
3	4	1	1	Futrelle, Mrs. Jacques Heath (Lily May Peel)	female	35.0	1	0	113803	53.1000	C123	S
4	5	0	3	Allen, Mr. William Henry	male	35.0	0	0	373450	8.0500	NaN	S

If the below is our dataset named data.csv

	Date	Time Worked	Money Earned
0	11/05/19	3	33.94
1	12/05/19	3	33.94
2	19/05/19	4	46.00
3	25/05/19	3	33.94
4	26/05/19	3	33.94

```
import pandas as pd
titanic_data = pd.read_csv('titanic.csv')
# Head diaplays inly top 5 rows from the data frame
titanic_data.head()
```

The following output is observed on successful execution of the above code.

```
      Date  Time Worked  Money Earned
0   11/05/19            3         33.94
1   12/05/19            3         33.94
2   19/05/19            4         46.00
3   25/05/19            3         33.94
4   26/05/19            3         33.94
```

And the output for displaying all the data sets-

```
df = pd.read_csv('titanic.csv')
print(df)
```

Output of the above lines

```
      Date  Time Worked  Money Earned
0   11/05/19            3         33.94
1   12/05/19            3         33.94
2   19/05/19            4         46.00
3   25/05/19            3         33.94
4   26/05/19            3         33.94
5    1/06/19            4         46.00
```

Adding more rows to the existing DataFrame (updating the rows of the DataFrame)

If the below is a different new.csv which we want to merge with the data.csv both having same attributes

```
      Date  Time Worked  Money Earned
0   10/06/19            3         33.94
1   12/06/19            4         46.00
2   14/06/19            3         33.94
```

```
df2 = pd.read_csv('new.csv')
df = df.append(df2, ignore_index=True, sort=False)
```

The output of the following code is

	Date	Time Worked	Money Earned
0	11/05/19	3	33.94
1	12/05/19	3	33.94
2	19/05/19	4	46.00
3	25/05/19	3	33.94
4	26/05/19	3	33.94
5	1/06/19	4	46.00
6	10/06/19	3	33.94
7	12/06/19	4	46.00
8	14/06/19	3	33.94

Example of some arithmetic operations that can be done on the dataset like summation-

```
Total_eatrnings = df['Money Earned'].sum()
Total_time = df['Time Worked'].sum()
print('You have earned total of ====>', round(Total_eatrnings),'CAD')
print('----------------------------------------')
print('You have worked fofr a total of ======>', Total_time, 'hours')
```

Here we added all the data of the attributes "Money Earned" and "Time Worked"

```
You have earned total of ====> 342 CAD
- - - - - - - - - - - - - - - - - - - - - - - -
You have worked for a total of ====> 30 hours
```

2.3   Plotting the Dataset in a Bar Graph

```
# [loting a bar graph using pandas library
df.plot(x='Date', y='Money Earned', kind = 'bar')
```

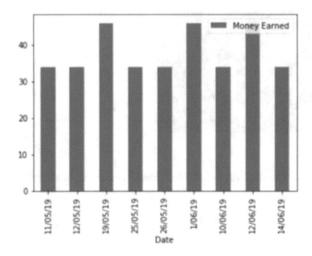

Reading selective data from data set

```
data = pd.read_csv('titanic.csv', skiprows=lambda x: x%2 !=0)
print(data)
```

The output is

```
     Date   Time Worked   Money Earned
0   12/05/19            3         33.94
1   25/05/19            3         33.94
2    1/06/19            4         46.00
```

Various other operations can be done in the same way with the help of lambda function.

2.4   Exception Handling

Handling exceptions for any random list

```
for entry in randomList:
    try:
        print("The entry is", entry)
        r = 1/int(entry)
        break
    except Exception as e:
        print("Oops!", e.__class__, "occurred.")
        print("Next entry.")
        print()
print("The reciprocal of", entry, "is", r)
```

Output:

```
The entry is a
Oops! <class 'ValueError'> occurred.
Next entry.

The entry is 0
Oops! <class 'ZeroDivisionError'> occurred.
Next entry.

The entry is 2
The reciprocal of 2 is 0.5
[Finished in 0.574s]
```

2.4.1   Try and Finally

This one ensures the file to be closed even after the error occurs.

```
try:
    f = open("test.txt",encoding = 'utf-8')
    # perform file operations
finally:
    f.close()
```

2.5   Quadratic Equation

```python
# Solving Quardratic Equation
from math import sqrt
print("In the equation aX^2 + bY^2 +c = 0 -")
a = input('Enter value of a : ')
b = imput('ENter value of b : ')
c = input('Enter value of c : ')
D = b*b - 4*a*b

if D > 0:
    roots = ( (-b + sqrt(D))/ 2*a, (-b - sqrt(D))/2*a )
    print(f'The roots of the eqution are real and un-equal')
    print(f'The roots value are : {roots[0]} and {roots[1]}')

elif D < 0:
    print('The Roots of the equations are irriational ')

else:
    roots = ( (-b + sqrt(D))/ 2*a, (-b - sqrt(D))/2*a )
    print(f'The roots of the eqution are real and un-equal')
    print(f'The roots value are : {roots[0]} and {roots[1]}')
```

Output:

```
In the equation aX^2 + bY^2 +c = 0 -
Enter value of a : 2
Enter value of b : 8
Enter value of c : 4
The roots of the eqution are real and un-equal
The roots value are : -8.0 and -8.0
```

2.5.1   Time and Date

```
>>> import datetime
>>> t = datetime.time(1, 10, 20, 13)
>>> print(t)
01:10:20.000013
>>> print(f'Hours:{t.hour} Minutes:{t.minute}, Seconds:{t.second}, Microsecond:{t.microsecond}')
Hours:1 Minutes:10, Seconds:20, Microsecond:13
>>> today = datetime.date.today()
>>> print(today)
2021-02-01
```

```
>>> print(today.strftime("%b %d %Y %H:%M:%S"))
Feb 01 2021 00:00:00
```

Matrix Operation:

```python
import numpy

# initializing matrices
x = numpy.array([[4, 2], [9, 3]])
y = numpy.array([[6, 8], [1, 10]])

# using add() to add matrices
print ("The element wise addition of matrix is : ")
print (numpy.add(x,y))

# using subtract() to subtract matrices
print ("The element wise subtraction of matrix is : ")
print (numpy.subtract(x,y))

# using divide() to divide matrices
print ("The element wise division of matrix is : ")
print (numpy.divide(x,y))
```

Output:

```
The element wise addition of matrix is :
[[10 10]
 [10 13]]
The element wise subtraction of matrix is :
[[-2 -6]
 [ 8 -7]]
The element wise division of matrix is :
[[0.66666667 0.25       ]
 [9.         0.3        ]]
```

Slicing in Row:

```python
import numpy as np
matrix_1 = np.array([np.arange(1,4), np.arange(4,7)])
print("Matrix: n", matrix_1)


# Slice to have 2nd row in matrix
print("result: n ", matrix_1[1:, :])
```

Output:

```
Matrix:
   [[1 2 3]
    [4 5 6]]
result:
   [[4 5 6]]
```

Slicing in Column:

```
import numpy as np
matrix_1 = np.array([np.arange(0,3), np.arange(3,6)])
print("Matrix: n", matrix_1)


# Slice to have 2nd column in matrix
print("result: n ", matrix_1[:, 2:])
```

Output:

```
Matrix:
   [[1 2 3]
    [4 5 6]]
result:
   [[3]
 [6]]
```

Slicing to Get Required Sub-matrix:

```
import numpy as np

matrix_1 = np.array([np.arange(1,5), np.arange(5,9), np.arange(9,13), np.arange(13,17)])
print("Matrix_1: n", matrix_1)
# Slice to get (2, 2) submatrix in the centre of mat_2d
print("sliced matrix: n ", matrix_1[1:3, 1:3])
```

Output:

```
Matrix_1:
 [[1 2 3 4]
 [5 6 7 8]
 [9 10 11 12]
 [13 14 15 16]]
sliced matrix:
        [[6 7]
         [10 11]]
```

For reader understanding this chapter mainly deals with RPi- and Arduino-based programming of sensors and relays.

Figure 2.1 elucidates the GPIO of the RPi zeroW.

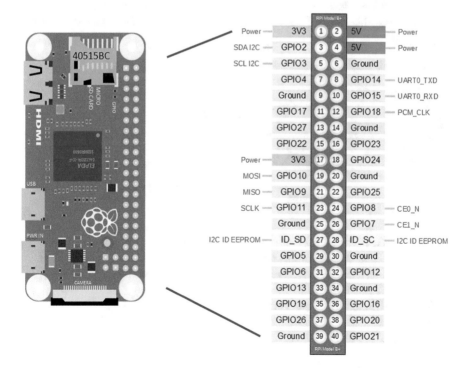

Fig. 2.1 Pin config of RPi. ZeroW

2.6   Python Programming

```
if loop
        num = 3
        ifnum> 0:
        print(num, "is a number.")

for loop
        for number in range(5):
        print("Thank you")

while loop
        number = 0
        while (number < 9):
        print 'The count is:', number
                number = number + 1

        print "thank you!"
```

Code 1 Loops in python

The python programming is compiled from the terminal of the raspberry pi. The command python filename.py is used to compile and run the program. The necessary packages, like RPi.GPIO, matplotlib, serial, etc., can be installed using pip installation

```
pip install RPi.GPIO

pip install pyserial
pip install matplotlib  etc.,
```

Code 2 Pip installation commands
  Code for RPi

```
import RPi.GPIO as GPIO    ##import GPIO library
import time             ## import time library
GPIO.setmode(GPIO.BOARD)  ## use board pin numbering
GPIO.setup(3, GPIO.OUT)   ## setup GPIO pin 7 to out
GPIO.output(3, True)      ## Turn on GPIO pin 7
time.sleep(10)
GPIO.output(3, False)
```

Code 3 RPi GPIO code

Figure 2.2 shows the circuit to connect led with the GPIO.

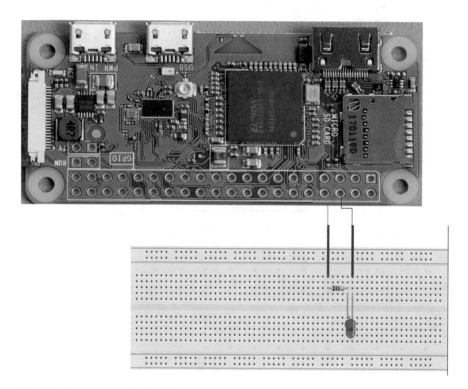

Fig. 2.2   Circuit connection for RPi

2.7   Precautions

The power pin and ground pin should not be short-circuited. Unknown value of resistor should not be used.

2.8   Relay Connections

Relays are used to handle high power loads and they can be controlled using the microcomputers like Arduino and RPi. The same code can be used to switch on and off the bulb connected with the relay unit. The bulb is connected with normally open terminal to switch on and off as shown in Figs. 2.3 and 2.4.

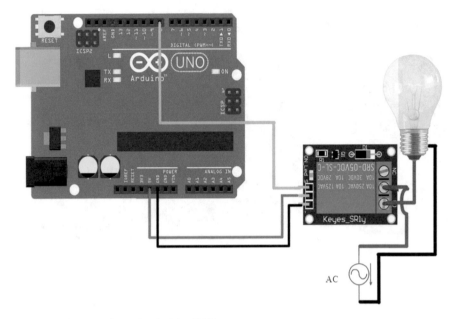

Fig. 2.3 Relay operation using Arduino UNO

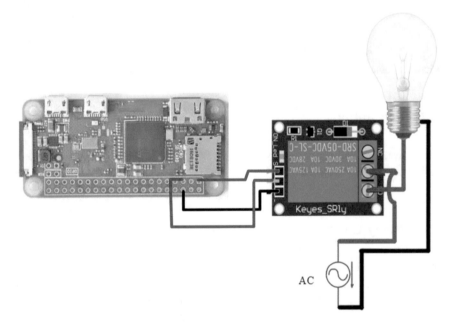

Fig. 2.4 Relay operation using RPi

2.9 Timer-Based Toggling

In case of timer-based toggling, the leds or relays are activated based on time delay. The time delay are created with time.sleep() and delay() commands in RPi and Arduino boards. Code 4 describes the arduino and raspberry pi codes for relay operation.

Arduino Program	Raspberry Pi
int pin = 12;	*import RPi.GPIO as GPIO*
void setup() {	*import time*
pinMode(pin, OUTPUT);	*GPIO.setmode(GPIO.BOARD*
digitalWrite(pin, HIGH);	*GPIO.setup(3, GPIO.OUT)*
delay(2000);	*GPIO.output(3,True)*
digitalWrite(pin, LOW)	*time.sleep(2000)*
}	*GPIO.output(3, False)*
void loop() {	
}	

Code 4 Delay programming in Arduino and Python
  Two seconds delay is set with delay() command

2.10 Sensor-Based Feedback Toggling

The acknowledgement of toggling of switch and sensors are done through LED glowing. That can be done by connecting switch and led (push button is considered as sensor here).
  Code 5 provides the push button-based relay operation for arduino and Rpi.

Arduino Program	Raspberry Pi
int push = 2;	import RPi.GPIO as GPIO
int bulb = 12;	import time
intbuttonState = 0;	GPIO.setmode(GPIO.BOARD)
void setup() {	GPIO.setup(3, GPIO.OUT)
pinMode(bulb, OUTPUT);	GPIO.setup(5, GPIO.IN)
pinMode(push, INPUT);	state = GPIO.input(pin)
}	while True:
void loop(){	if (state is True)
buttonState = digitalRead(push);	GPIO.output(3,True)
if (buttonState == HIGH)	else
{	GPIO.output(3, False)
digitalWrite(bulb, HIGH);	GPIO.cleanup()
}	
else	
{	
digitalWrite(bulb, LOW);	
}	
}	

Code 5 Switch programming in Arduino and RPi

2.11 Interrupt-Based Toggling

When the microcomputer (Arduino or RPi) receives any interrupt signal, its hardware interrupt program is enabled. Though the Microcomputer is processing some other process. This interrupt is given priority and an interrupt subroutine is executed during each interrupt detection. The circuit diagram for the hardware interrupt programming in case of Arduino Uno and RPi are given in Fig. 2.5. Code 6 shows the arduino- and RPi-based interrupt programming.

Arduino Program	Raspberry Pi
int pin = 2;	import RPi.GPIO as GPIO
volatile int state = LOW;	GPIO.setmode(GPIO.BCM)
void setup() {	GPIO.setup(23, GPIO.IN, pull_up_down=GPIO.PUD_UP)
pinMode(12, OUTPUT);	raw_input("Press Enter when ready\n")
attachInterrupt(digitalPinToInterrupt(pin), blink, CHANGE);	print "Press your button when ready to initiate a falling edge interrupt."
}	try:
void loop() {	GPIO.wait_for_edge(23, GPIO.FALLING)
digitalWrite(12, state);	print "\nFalling edge detected. Now your program can continue with"
}	except KeyboardInterrupt:
	GPIO.cleanup()
void blink() {	GPIO.cleanup()
state = !state;	
}	

Code 6 Interrupt programming in Arduino and Python

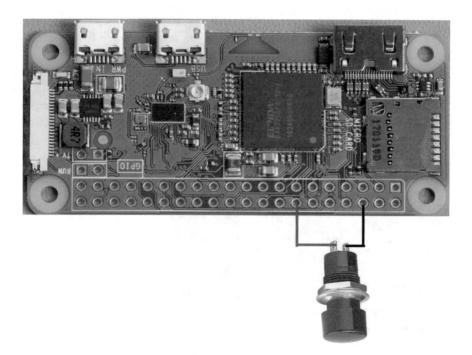

Fig. 2.5 Hardware Interrupt toggling

The USB camera is attached with the RPi and the following package is installed using wget command in terminal window. Raspberry pi is 32 bit processor and can operate and handle camera data at faster rate. This section deals with RPicam programming with Python

wget http://www.cl.cam.ac.uk/projects/raspberrypi/tutorials/robot/resources/RPi.GPIO-0.3.1a.zip

unzip RPi.GPIO-0.3.1a.zip

cd RPi.GPIO-0.3.1a

sudo python setup.py install

cd ..

wget http://www.cl.cam.ac.uk/projects/raspberrypi/tutorials/robot/resources/i2c.py

sudo mv i2c.py /usr/local/lib/python2.7/dist-packages/i2c.py

wget http://www.cl.cam.ac.uk/projects/raspberrypi/tutorials/robot/resources/imgproc.zip

unzip imgproc.zip

cd library

sudo make install

cd ..

After installing the imgproc package, the following code can be executed to take a photo from the RPi. Code 7 provides the image processing camera access with Python.

```
import imgproc

import time

from imgproc import *

camera = Camera(160, 120)

viewer = Viewer(320, 240, "The Image")

img = camera.grabImage()

time.sleep(100)

viewer.displayImage(img)
```

Code 7 Camera access with imgproc library

If the pi camera is used, a package called picamera array is used. The picamera array package can be installed through executing pip command in the terminal window.

pip install "picamera[array]"

```
from picamera.array import PiRGBArray

from picamera import PiCamera

import time

import cv2

camera = PiCamera()

rawCapture = PiRGBArray(camera)

time.sleep(0.1)

camera.capture(rawCapture, format="bgr")

image = rawCapture.array

cv2.imshow("Image", image)

cv2.waitKey(0)
```

Code 8 Camera image capture with cv2 library

For video streaming in RPi, the same code with while loop is executed as given in code 9.

```
import imgproc

import time

from imgproc import *

camera = Camera(160, 120)

viewer = Viewer(320, 240, "The Image")

img = camera.grabImage()

time.sleep(100)

while (True):

        img = camera.grabImage()

        viewer.displayImage(img)
```

Code 9 Video Streaming in RPi with Python

The acquitted image from camera sensor in RPi can be saved as JPG file with the following code 10.

```
import cv2
camera = cv2.VideoCapture(0)
while True:
return_value,image = camera.read()
gray = cv2.cvtColor(image,cv2.COLOR_BGR2GRAY)
cv2.imshow('image',gray)
if cv2.waitKey(1)& 0xFF == ord('s'):
        cv2.imwrite('test.jpg',image)
        break
camera.release()
cv2.destroyAllWindows()
```

Code 10 Saving capture file in jpg format in CV2

The edge detection in RPi can be done through the following code 11.

```
import imgproc
from imgproc import *
import math
camera = Camera(160, 120)
viewer = Viewer(160, 120, "The image")
camera.grabImage()
camera.grabImage()
hig = camera.grabImage()
img_edge = Image(160, 120)
viewer.displayImage(img)
for x in range(1, hig.width - 1):
        for y in range(1, hig.height - 1):
                Gx = 0
                Gy = 0
                r, g, b = hig[x - 1, y - 1]
                intensity = (r + g + b)
                Gx += -intensity
                Gy += -intensity
                r, g, b = hig[x - 1, y]
                Gx += -2 * (r + g + b)
                r, g, b = hig[x - 1, y + 1]
                Gx += -(r + g + b)
                Gy += (r + g + b)
                r, g, b = hig[x, y - 1]
                Gy += -2 * (r + g + b)
                r, g, b = hig[x, y + 1]
                Gy += 2 * (r + g + b)
                r, g, b = hig[x + 1, y - 1]
                Gx += (r + g + b)
                Gy += -(r + g + b)
                r, g, b = hig[x + 1, y]
                Gx += 2 * (r + g + b)
                r, g, b= hig[x + 1, y + 1]
                Gx += (r + g + b)
                Gy += (r + g + b)
                length = math.sqrt((Gx * Gx) + (Gy * Gy))
                length = length / 4328 * 255
                length = int(length)
                img_edge[x, y] = length, length, length
viewer.displayImage(img_edge)
waitTime(5000)
```

Code 11 Edge detection using Python.

2.12 Text to Speech

The chatbot and other public interacting bots mainly use Text to Speech(TTS) library. The TTS library can be tuned with different voice and accent, the TTS upgradation is also available for languages like French, Hindi, English, etc. Text to

Speech, the following python code converts the input text to voice. The following code is executed in terminal window to install python tts package.

sudo pip install pyttsx

The following code makes the RPi to speak the typed text, the headphones are to be connected before executing the code 12.

```
import pyttsx
eng = pyttsx.init()
eng.setProperty('rate', 70)
voices = eng.getProperty('voices')
for voice in voices:
print "Using voice:", repr(voice)
eng.setProperty('voice', voice.id)
eng.say("Hi welcome to IoT lab?")
eng.say("A B C D E F G H I J K L M")
eng.say("N O P Q R S T U V W X Y Z")
eng.say("0 1 2 3 4 5 6 7 8 9")
eng.say("Sunday Monday Tuesday Wednesday Thursday Friday Saturday")
eng.runAndWait()
```

Code 12 Text to Speech code in Python

The following code 13 captures the wordings in image and provides as a string.

```
import pyttsx
from PIL import Image
from pytesseract import image_to_string
eng = pyttsx.init()
eng.setProperty('rate', 70)
myText = image_to_string(Image.open("tmp/test.jpg"),config='-psm 10')
myText = image_to_string(Image.open("tmp/test.jpg"))
voices = eng.getProperty('voices')
eng.setProperty('voice', voice.id)
eng.say(myText)
eng.runAndWait()
```

The following code can be used to update and values in the Google sheet using gspread package. The gspread package can be installed by executing pip install gspread in the terminal window of the RPi. The following code enters the value 1, 2, 3, and 4 in the google sheet cells. The json download and other preliminary activities required for getting access with the gspread sheet is given in link https://gspread.readthedocs.io/en/latest/.

pip install gspread

Code 13 Image to String conversion Python

A spreadsheet named singlemachine is opened in Google sheet and its values were read using code 14.

```
import sys

import gspread

from oauth2client.service_account import ServiceAccountCredentials

import urllib

scope = ['https://spreadsheets.google.com/feeds']

creds =
ServiceAccountCredentials.from_json_keyfile_name('singlemachine-
0319fe351cb6.json', scope)

client = gspread.authorize(creds)

sheet = client.open("singlemachine").sheet1

sheet.cell(1, 1).value

list_of_hashes = sheet.cell(1, 1).value

sheet.update_cell(1, 1, 1)

sheet.update_cell(2, 1, 2)

sheet.update_cell(3, 1, 3)

sheet.update_cell(4, 1, 4)
```

Code 14 Gspread data access with Python

The following code 15 can be used to interface rfid and retrieve the necessary details from the database. The MySQLdb is used as a database for storing information in tables. The code authenticates and opens the door for the authorized employees only. The door is connected with relay connections and motors.

```
import serial
importMySQLdb
import string
importRPi.GPIO as GPIO
import time
conn=MySQLdb.connect(host="local host", user="root", passwd="root",
b="vehicles")
cursor=conn.cursor()
print "database connected"
GPIO.setmode(GPIO.BCM)
GPIO.setup(10, GPIO.OUT)
GPIO.setup(4, GPIO.OUT)
ser = serial.Serial(" /dev/ttyAMA0" , 9600)
ser.close()
ser.open()
p= ser.readline()
print p
c=p
cursor.execute("SELECT * FROM rf0012803899")
rows = cursor.fetchall()
print rows
GPIO.output(10, GPIO.HIGH)
GPIO.output(4, GPIO.HIGH)
time.sleep(15)
GPIO.output(10, GPIO.LOW)
GPIO.output(4, GPIO.LOW)
ser.close()
```

Code 15 Python Code for RFID data access with serial Python

Figure 2.6 provides the RFID-based door operation using RPi ZeroW.

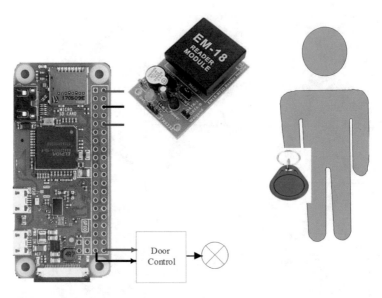

Fig. 2.6 Door authentication for employee using RFID

The sensor value recording in RPi is stored in data base with time stamp. The same can be used for analyzing distributions and observing frequency of occurrences.

References

1. *Internet of Things and Additive Manufacturing: towards an* (New York, Springer, 2020). https://www.springer.com/gp/book/9783030325299
2. S. Rani, R. Maheswar, G.R. Kanagachidambaresan, P. Jayarajan, *Integration of WSN and IoT for smart cities* (Springer, Cham, 2020)
3. M. Lutz, *Learning Python* (O'Reilly Media, Beijing, 2013)
4. M. Lutz, *Programming python*, 2nd edn. (O'Reilly, Beijing, 2001). https://link.springer.com/book/10.1007%2F978-3-030-57077-4
5. J.M. Zelle, *Python programming: an introduction to computer science* (Franklin, Beedle & Associates, Incorporated, Wilsonville, OR, 2004)
6. *Programming with tensorflow: solution for edge computing applications.* S.l.: Springer (2020)

Chapter 3
Sensors and SBCs for Smart City Infrastructure

3.1 Introduction

Sensors are the front end data collection devices which converts the physical events to electrical signals and further processed by adc converter and processing unit in smart city applications [1–21]. The sensors are basically classified based on the operation and their outputs. The sensors are basically (Fig. 3.1)

- Colorimetric
- Voltammetry

© The Author(s), under exclusive license to Springer Nature Switzerland AG 2021
G. R. Kanagachidambaresan, *Role of Single Board Computers (SBCs) in rapid IoT Prototyping*, Internet of Things, https://doi.org/10.1007/978-3-030-72957-8_3

| Moisture | Flex | Touch | Solar light | Metal detection |

| RTC | Vibration | Thermistor | Infrared | Ultrasonic |

| Gyro | Accelero meter | Color | PIR | GAS |

| Smoke | Temperature | LDR | Rainfall | Soil Moisture |

Fig. 3.1 Most commonly used arduino-based sensors

Table 3.1 provides the sensors specifications with respect to its applications in various important fields [22–25].

Table 3.1 Comparison table of sensors with its application field

Field	Sensor name	Image	Working Principle	Monitoring parameter	Operating voltage (current)	Frequency of operation	Communicating protocol	Sensing range	Operating rage	Recovery time
Agriculture	Soil temperature		The soil moisture sensor is a device to measure volumetric water content in the soil.	1. These sensors measure the water content of a soil using the time or frequency of a pulse traveling between or returning to electrodes. The most common types are capacitance and time or frequency domain. 2. Most sensors are accurate within 2–3% of the actual soil moisture	3.3–5 V	70 MHz	Digital value output connector(0 or 1)	In the range of ADC value from 0 to 1023.	Measuring range temperature 0–60 °C	Once the rode taken out it can be measured again in another place
	Depth sensor (water level)		This water level sensor module has a series of parallel exposed traces to measure droplets/ water volume in order to determine the water level	Very Easy to monitor water level as the output to analog signal is directly proportional to the water level. This output analog values can be directly read via ADC and can also be connected directly Arduino's analog input pins.	DC 3–5 V	Analog	The Sensor is the Analog type which produces analog output signals according to the water pressure with its Detection Area	10–90% non-condensing	Operating Temperature (°C): −10–30	It depends on the moisture level
	Air temperature		The air temperature sensor is a thermistor, which means its electrical resistance changes in response to changes in temperature.	It works the same as a coolant sensor. The PCM applies a reference voltage to the sensor (usually 5 volts), then looks at the voltage signal it receives back to calculate air temperature.	5 V	900 MHz	Communicate data directly to the RX3000 weather station	±0.2°C (±0.36°F) and ±2.5% RH	from −40 °C to +120 °C	

(continued)

Table 3.1 (continued)

Rain sensor		Rain Sensor or detector is simple and easy to use the module for rain detector. The module works as a switch when rain falls on the module and also measures rain intensity	This rain module can measure beyond humidity detector, which means it can measure what humidity cannot	5 V	100 kΩ to 2 MΩ	I2c	large area of 5.5 * 4.0 CM	High durability	Once it completely dry from moisture
Leaf wetness sensor		It measures leaf wetness by determining the electrical resistance on the surface of the sensor	It is primarily used to determine the percentage of time that a leaf surface is wet, versus the time it is dry	5 V	transition is between 50 and 200 kohm	I2c	$-40\ ^\circ C$ to $+150\ ^\circ C$ Sensor may crack when temperature drops below $-40\ ^\circ C$.	0 °C to 100°C	Droplets must touch two fingers simultaneously to change the sensor resistance
Chlorophyll sensor		The probe is designed to estimate phytoplankton concentrations by detecting the fluorescence from chlorophyll in situ.	Dissolved oxygen, turbidity, chlorophyll, blue-green algae, and/or rhodamine), along with other parameters	12.0 V battery		SDI -12 communication	60m	• Detection Limit: ~0.1 μg/L	
Wind direction sensor		Wind vanes measure wind direction and are often used with anemometers, which measure wind speed.	Anemometers measure wind speed and wind vanes measure wind direction. A typical wind vane has a pointer in front and fins in back. When the wind is blowing, the wind vane points into the wind.	4–5 V	0–60 mph	• In areas where icing of the anemometer is a problem, use Anemometer Drip Rings to deflect water from the joint between moving parts	-40 to $+50\ ^\circ C$ (equivalent to 305 to 368 m s-1 in speed of sound)	• 0–30 m/s	Needs 10 s to restart

(continued)

Field	Sensor name	Image	Working Principle	Monitoring parameter	Operating voltage (current)	Frequency of operation	Communicating protocol	Sensing range	Operating rage	Recovery time
	Solar radiation		Sensor chip is suitable for detecting the UV radiation in sunlight.	provides at its output a voltage proportional to the intensity of the light in the visible range of the spectrum,	2.5–5 V		I2c	240 nm to 370 nm	• 130°	
	Pressure sensor		It includes pressure measurement in weather networks, often for weather forecasting and to correct the output of sensors that are sensitive to pressure changes.	A pressure sensor works by converting pressure into an analog electrical signal. The demand for pressure measuring instruments increased during the steam age	12–24 V	80 Hz	IP 65	0–115 kpa	• 40–80 °C	50 k pa
	PH sensor		The sensor used to measure pH of a liquid or solid substances is known as pH sensor	The Analog pH Sensor Kit is specially designed for Arduino controllers and has a built-in simple, convenient, and practical connection	5 V DC.		I2c	0–14 PH	±0.01 pH	≤1 min
	Potassium NITRATE SENSORS		Potassium Sensor converts the analog signals from the electrode cartridge into a digital protocol	1. An optical transducer is developed to measure and to detect the presence of Nitrogen (N), Phosphorus (P) and Potassium (K) of soil. 2)The nutrient absorbs the light from LED and the photodiode convert the remaining light that is reflected by reflector to current.	5 mV		allows two way communications with the transmitter	Potassium :20 ppb to 39,000ppm Ph :2.5 to 11ph	Tem range: 0 °C to 40 °C	T90 in 10 s

Table 3.1 (continued)

Field	Sensor name	Image	Working Principle	Monitoring parameter	Operating voltage (current)	Frequency of operation	Communicating protocol	Sensing range	Operating rage	Recovery time
	Relative humidity		A humidity sensor is an electronic device that measures the humidity in its environment	The electrodes are placed in interdigitized pattern to increase the contact area. The resistivity between the electrodes changes when the top layer absorbs water and this change can be measured with the help of a simple electric circuit	3.3–5 V	200 Hz and 10 kHz	DHT11/DHT22	Humidity 20–90%RH, Temperature 0–50°C	<±1%RH/year 50%	recovery time of 47 s
Smart trans-portation	GPS sensor		This is a complete GPS module that is based on the Ublox NEO-6M. This unit uses the latest technology from Ublox to give the best possible positioning	The GPS receiver gets a signal from each GPS satellite. The satellites transmit the exact time the signals are sent. … The GPS receiver also knows the exact position in the sky of the satellites, at the moment they sent their signals.	3.3 OR 5 V	1.1–1.5 GHz	ip	within a 4.9 m (16 ft.)	L1 (1575.42 MHz) and L2 (1227.60 MHz)	Once the internet is connected to the device
	Ultrasonic		This ultrasonic sensor module can be used for measuring distance, object sensor, motion sensors etc.	The module sends eight 40 kHz square wave pulses and automatically detects whether it receives the returning signal. If there is a signal returning, a high-level pulse is sent on the echo pin..	5 V	40 kHz to 70 kHz	Serial communication	2–4 m	11 m	Distance of the object

Lidar laser		providing accurate distance measurement whatever the target reflectances, unlike conventional technologies. It can measure absolute distances up to 2m, setting a new benchmark in ranging performance levels	A typical lidar sensor emits pulsed light waves into the surrounding environment. These pulses bounce off surrounding objects and return to the sensor. The sensor uses the time it took for each pulse to return to the sensor to calculate the distance it traveled.	2.6 to 5.5	10 GHz	I2C	940 nm	2 m	Up to 15 m
Stereovision sensors		A stereo-based vision sensor was developed, together with the design of an algorithm for target sensing. Target sensing means outputting the distance and direction of the vehicle	This binocular stereo vision expansion board is specially designed for Raspberry Pi Compute Module, compatible with CM3 / CM3 Lite / CM3+ / CM3+ Lite. It features three camera ports and commonly used ports like DSI and USB, more other peripheral interfaces are also supported through the FPC connector, all in the small size body	CR1220 battery holder	7 MHz	I2C	−40 °C to +85 °C		

(continued)

Table 3.1 (continued)

Field	Sensor name	Image	Working Principle	Monitoring parameter	Operating voltage (current)	Frequency of operation	Communicating protocol	Sensing range	Operating rage	Recovery time
Smart health care sensor	Pulse rate sensor		Very small size a plug-and-play heart rate sensor for Arduino and Arduino compatible boards. It can be used by students, artists, athletes, makers, and game and mobile developers who want to easily incorporate live heart-rate data into their projects.	Pulse Sensor Amped adds amplification and noise cancelation circuitry to the hardware. It is noticeably faster and easier to get reliable pulse readings.	3 V or 5 V	20 Hz		0.5″x0. 7	The Heart Rate sensor measures heart rate between 0 and 250 bpm	23 beats per minute
	ECG sensor		MAX30003 is a single-lead ECG monitoring IC which has built-in R-R detection	ECG records the electrical activity generated by heart muscle depolarizations, which propagate in pulsating electrical waves towards the skin. ECG electrodes are typically wet sensors, requiring the use of a conductive gel to increase conductivity between skin and electrodes.	1.8 V	125 Hz	I2C	25 mm/s	0.08 to 0.10 s (80–100 ms) in duration.	

EEG sensor		EEG is a small electrodes and wires are attached to your head. The electrodes detect your brain waves and the EEG machine amplifies the signals and records them in a wave pattern on graph paper or a computer screen	Electroencephalography (EEG) is a method to record brain activity throw the capture of electric activation. This neurophysiological measurement can be acquired by non-invasive scalp electrodes. The measurement is the summation of post-synaptic neuron potentials within a large area (1 to 6 cm2) of the cortex.	microvolts (mV)	Delta: has a frequency of 3 Hz Theta: has a frequency of 3.5 to 7.5 Hz Alpha: has a frequency between 7.5 and 13 Hz Beta: beta activity 14 and greater Hz.	10 MM around head	(1 to 6 cm2) of the cortex
EKG sensor		EKG tracings to record electrical activity in the heart	During an ECG , up to 12 sensors (electrodes) will be attached to your chest and limbs. The electrodes are sticky patches with wires that connect to a monitor. They record the electrical signals that make your heart beat. A computer records the information and displays it as waves on a monitor or on paper	• Offset: ~1.00 V (±0.3 V) • Gain: 1 mV body potential / 1 V sensor output	0.01–250 Hz	Normal range 120 – 200 ms	420 ms

(continued)

Table 3.1 (continued)

Field	Sensor name	Image	Working Principle	Monitoring parameter	Operating voltage (current)	Frequency of operation	Communicating protocol	Sensing range	Operating rage	Recovery time
	Temperature sensor		This sensor module is an infrared thermometer for noncontact temperature measurements.	MLX90614 is an infrared non-contact thermometer, and the TO-39 package integrates infrared induction thermoelectric pile detector chip (MLX81101) and signal processing dedicated integrated chip MLX90302. The sensor achieves high precision and high-resolution measurement.	3.3–5V	546 Hz/ o C	Up to 127 sensors can be read via common 2 wires	25–75 cm	−40 °C to +125 °C	1 s
	BP sensor		A digital blood pressure monitor uses an air pump to inflate a cuff surrounding an upper arm or a wrist with sufficient pressure to prevent blood flow in the local main artery	This pressure cuff linked to a mercury column is used to measure the blood pressure. Here, the doctor manually pumps the cuff to increase the pressure on the artery. Then using stethoscope the noise of the blood rushing through the artery.	3.3 to 5 V	3.4 MHz	I2C	300 hPa to 1100 hPa (+9000 m to −500 m)	0 to 375 mmHg	5 ms
Smart city sensor	Ultrasonic		This ultrasonic sensor module can be used for measuring distance, object sensor, motion sensors etc.	The module sends eight 40 kHz square wave pulses and automatically detects whether it receives the returning signal. If there is a signal returning, a high-level pulse is sent on the echo pin.	5 V	40 kHz to 70 kHz	Serial communication	2–4 m	11 m	Distance of the object

Sensor		Description					
Radiation level sensor		Sensor chip is suitable for detecting the UV radiation in sunlight.	provides at its output a voltage proportional to the intensity of the light in the visible range of the spectrum,	2.5–5 V		I2c	240–370 nm	• 130°
Air pollution sensor		Electrochemical sensors are based on a chemical reaction between gases in the air and the electrode in a liquid inside a sensor	Optical sensors detect gases like carbon monoxide and carbon dioxide by measuring the absorption of infrared light	10 V		DHT 11	0–1000 ppm	• 0 –25 c <30 s
Current sensor		ACS712 is a current sensor module which is compatible to sense both AC and DC current	This sensor uses the principle of hall effect to measure the current.	5 V	120 MHz		30 A	• Both AC and DC
Voltage sensor		A voltage sensor is a sensor is used to calculate and monitor the amount of voltage in an object. Voltage sensors can determine both the AC voltage and DC voltage level.	It is based on principle of resistive voltage divider design. It can make the red terminal connector input voltage to 5 times smaller. Arduino analog input voltages up to 5V, the voltage detection module input voltage not greater than 5Vx5=25V	25 V	50 Hz		Detection range: DC 0.02445–25 V	• Input voltage range: 0.025–25 V max.

(continued)

Table 3.1 (continued)

Field	Sensor name	Image	Working Principle	Monitoring parameter	Operating voltage (current)	Frequency of operation	Communicating protocol	Sensing range	Operating rage	Recovery time
	Vibration sensor		It detect if there is any vibration that beyond the threshold. The threshold can adjust using an onboard potentiometer. When this no vibration, this module output logic LOW the signal indicates LED light, and vice versa.	The vibration sensor is also called a piezoelectric sensor. These sensors are flexible devices which are used for measuring various processes. This sensor uses the piezoelectric effects while measuring the changes within acceleration, pressure, temperature, force otherwise strain by changing to an electrical charge.	5 V	sensitivity 100 mV/g,		Can work on low voltage	• 0°C to +80 °C	2 ms
	Piezoelectric sensor		Piezoelectricity is the electric charge that gathers when the mechanical stress is applied in the solid materials.	The piezoelectric effect is utilized to measure changes in vibration, acceleration, pressure, strain, or force by changing them into electrical energy. It is utilized to detect the changes of a motor parameter like speed, velocity, temperature.		1.75 kHz + 0.1 kHz		40 mm	• -30 °C –+70 °C	
Defense	Infrared sensor		an infrared (IR) sensor is an electronic device that measures and detects infrared radiation in its surrounding environment	It does this by either emitting or detecting infrared radiation. Infrared sensors are also capable of measuring the heat being emitted by an object and detecting motion.	5 V		I2C	1 cm	• -40 to + 125 °C	

Category	Name	Image	Description	Voltage	Frequency	Notes	Dimensions	Range	Time
	Swir cameras		Short Wave Infrared (SWIR) cameras are used in astronomy research to study	This short wave infrared sensor study the J-, H-, and K- bands. They offer high resolution, and are cooled which makes them ideal for telescopes. Integration times can be as long as 3 minutes with low readout noise when cooled to -60°C.	12 V	Up to 100 Hz		9.6 mm × 7.68 mm	• Low Gain: 62 dB High Gain: 56 dB	- 300 s
	Motion detection sensor		The Passive Infrared Sensor (PIR) sensor module is used for motion detection	Active motion sensors emit ultrasonic sound waves that reflect off objects and bounce back to the original emission point. When a moving object disrupts the waves, the sensor triggers and completes the desired action, whether this is switching on a light or sounding an alarm	4.4–20 V	24 GHz		<140°	• to 7 m	5 to 200 s
Water manage-ment	Temperature sensor		This sensor module is an infrared thermometer for noncontact temperature measurements.	MLX90614 is an infrared non-contact thermometer, and the TO-39 package integrates infrared induction thermoelectric pile detector chip (MLX81101) and signal processing dedicated integrated chip MLX90302. The sensor achieves high precision and high-resolution measurement.	3.3–5 V	546 Hz/ o C	- up to 127 sensors can be read via common 2 wires	25–75 cm	−40 to +125 °C	1 s

(continued)

Table 3.1 (continued)

Field	Sensor name	Image	Working Principle	Monitoring parameter	Operating voltage (current)	Frequency of operation	Communicating protocol	Sensing range	Operating rage	Recovery time
	PH sensors		A pH sensor is one of the most essential tools that is typically used for water measurements.	This type of sensor is able to measure the amount of alkalinity and acidity in water and other solutions. In most cases, the standard pH scale is represented by a value that can range from 0-14	5 V		I2C	0–14 PH	±0.01 pH	1 min
	Turbidity Sensor		Turbidity meter for the detection of turbidity of liquids and aqueous solutions can be found here Turbidity sensors measure the amount of light that is scattered by the suspended solids in water.	A turbidity probe works by sending a light beam into the water to be tested. This light will then be scattered by any suspended particles. The amount of light reflected is used to determine the particle density within the water. The more light that is detected, the more particles are present in the water	2.1–3.6 V	Rs232	SHT20	100M	5–90 °C	<500 ms
	Conductivity sensor		The conductivity is measured by applying an alternating electrical current to the sensor electrodes (that together make up the cell constant) immersed in a solution and measuring the resulting voltage. The solution acts as the electrical conductor between the sensor electrodes.	A conductivity sensor (or conductivity probe) measures the ability of a solution to conduct an electrical current. It is the presence of ions in a solution that allow the solution to be conductive: the greater the concentration of ions, the greater the conductivity.	24 V		1 kHz	500 ft	−5 °C to +70 °C.	10 s

Sensor	Image	Description		Supply	Frequency	Range		Response time
Optical do sensor		Optical DO sensors, popularly known as luminescent DO sensors (LDO) but some are called fluorescent sensors, measure dissolved oxygen concentration in water based on the quenching of luminescence in the presence of oxygen	They can measure either the intensity or the lifetime of the luminescence as oxygen affects both.	5 V	20 kHz to 200 kHz	0–20 mg/L 1 mm	1 mm	20–30 s
Fluorometer sensor		Fluorescence occurs when one wavelength of light hits and excites electrons in a material, and the electrons in that object instantaneously emit (fluoresce) light of a different wavelength	A fluorometer is a device that measures the fluorescence or light emitted by different fluorescing objects.	5 V		Pressure : 600 dbar	–2–50 °C	
Ion-selective electrodes		This sensor is a combination ion-selective electrode (ISE) for the determination of fluoride (F⁻) in solution.	The lanthanum fluoride crystal membrane produces a potential change due to the fluoride ion exchange between the membrane and the sample. The internal sensing elements are housed within a durable body, composed of a mixture of polyetherimide (PEI) and epoxy.	5 V		0.02 mg/L	Detection from 0.62 to 6200 mg/L NO_3	

(continued)

Table 3.1 (continued)

Field	Sensor name	Image	Working Principle	Monitoring parameter	Operating voltage (current)	Frequency of operation	Communicating protocol	Sensing range	Operating rage	Recovery time
	Total dissolved gas sensors		Total Dissolved Gas (TDG) sensor uses a pressure transducer mounted behind a rigid gas-permeable membrane to measure amount of total gaseous compounds dissolved in a liquid.	Real-time measurement indicates water supersaturated with atmospheric gases, which can cause gas bubble gill disease in aquatic organisms	2.5 V			0–50 mg/L	1.0 mm zHg	<2 s
Outer Space Applications	Sound sensor		A sound sensor is defined as a module that detects sound waves through its intensity and converting it to electrical signals.	Sound is detected via microphone and fed into an LM393 op amp. The sound level set point is adjusted via an on board potentiometer. When the sound level exceeds the set point, an LED on the module is illuminated and the output is sent low.	5 V		16–20 kHz	52–48 dB	Noise level 3–6 kHz	10 s
	Temperature sensor		This sensor module is an infrared thermometer for noncontact temperature measurements.	MLX90614 is an infrared non-contact thermometer, and the TO-39 package integrates infrared induction thermoelectric pile detector chip (MLX81101) and signal processing dedicated integrated chip MLX90302.the sensor achieves high precision and high-resolution measurement.	3.3–5 V	546 Hz / o C	Up to 127 sensors can be read via common 2 wires	25–75 cm	−40 °C to +125 °C	1 s

Sensor		Description							
Radiation Sensor		A radiation detector is a device for measuring nuclear, electromagnetic or light radiation	A nuclear radiation detector identifies nuclear radiation by measuring the emission of ionizing radiation of alpha particles, beta particles and gamma rays	5 V			100 mm²	100 mm²	15 s
Radar motion detector sensor		These Radar sensors are used for detecting, locating, tracking, and recognizing objects of various kinds at considerable distances	It operates by transmitting electromagnetic energy towards objects, commonly referred to as targets, and observing the echoes returned from them.	5 V		5.8 GHz	20 m	180°	1.6 s
Spectrometer sensor		Spectrometers have internal sensors that can instantaneously measure the light and divide the incoming signal across a detector array which measures the signal in small bands or individual wavelengths based on the resolution of the system	A spectrometer is a measuring device that collects light waves. … When objects are hot enough, they emit visible light at a given point or points on the electromagnetic spectrum. Spectrometers split the incoming light wave into its component colors. Using this, they can determine what material created the light	5 V	I2C	18 frequencies of light-sensing from 410 nm to 940 nm.			60 s

Most of the sensors for IoT are connected with I2C and SPI communication protocol. The i2c and SPI communication protocol in arduino does not require enabling. However in raspberry pi and other similar boards the GPIOs needs to be enabled with config commands. The enabling of i2c and spi GPIO pins is given in the following figures. Figure 3.2 elucidates the interfacing option in raspberry pi. The following command can be use to config the Rpi.

sudoraspi-config

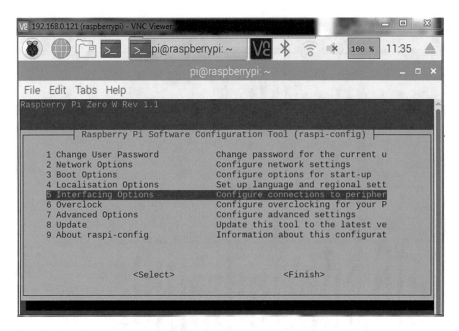

Fig. 3.2 Raspiconfig command screen 1

Once the i2c port is enabled with the sudo-raspiconfig command. The availability of the devices connected with the i2c ports can be identified with the following command

i2cdetect –y 1

Figure 3.3 illustrates the output screenshot of the rpi command window after the execution of i2cdetect command.

Fig. 3.3 i2c devices commands connected

Figure 3.4 elucidates the devices connected in the i2c ports; the device is connected in the address 0x48.

Fig. 3.4 Address of i2c connected with raspberry pi

Once the two devices are connected with i2c devices the address matrix is updated, one device is connected with 0x48 and 0x70 as given in Fig. 3.5.

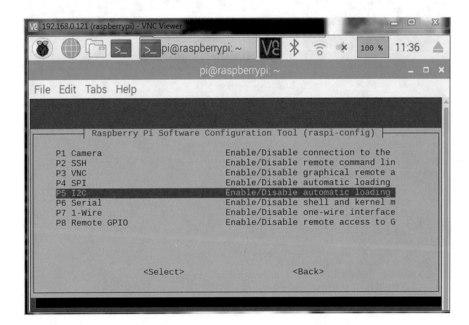

Fig. 3.5 Address matrix with two i2c devices connected

If the devices are not connected and address identification is not shown, the i2c communication port has to be enabled with the raspi-config command in the terminal. The default OS does not allow i2c, spi, camera, and sshconnectivities; hence it has to be enabled with raspi-config command. Figure 3.6 elucidates the raspi-config command option in terminal.

Fig. 3.6 i2c communication port activation using raspi-config command

Python code for i2c connected ads1115 for analog to digital conversion. Figure 3.7 elucidates the python code to receive channel values of the ads1115 analog to digital converter. Figure 3.7 elucidates the python program for the ads1115 converter. The raspberry pi on default does not have analog pins like arduino; hence this section focuses on widely used ads1115 ic.

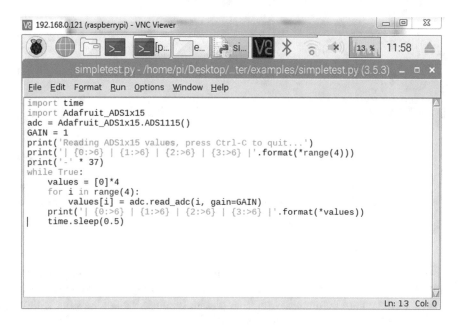

Fig. 3.7 ADS1115 python code for raspberry pi

Once the module is designed, the raspberry pi is powered and analog sensors are connected to the ports. The python program is executed in terminal window of the raspberry pi, and the corresponding digital values are seen in the terminal window. Figure 3.8 elucidates the analog to digital value read from the ads1115. The circuit and interfacing with the IC is discussed in upcoming chapters.

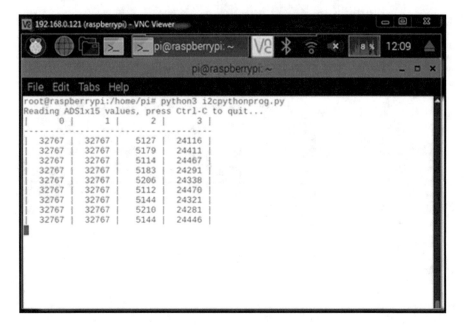

Fig. 3.8 Analog values of 4 channel ads1115

The ads1115 ic can be able to read 4 channel with 16 bit precision; in case more sensors has to be connected with the raspberry pi, an 10 bit precision mcp3008 ic which works on spi communication is available. Figure 3.9 elucidates the python program to access the mcp3008 ic. The MCP3008 IC can accommodate 8 channels and can connect analog to 10 bit precision digital value.

```
import time
importspidev

spi_ch = 0

# Enable SPI
spi = spidev.SpiDev(0, spi_ch)
spi.max_speed_hz = 1200000

defread_adc(adc_ch, vref = 3.3):

    # Make sure ADC channel is 0 or 1
ifadc_ch != 0:
adc_ch = 1

    # Construct SPI message
# First bit (Start): Logic high (1)
# Second bit (SGL/DIFF): 1 to select single mode
# Third bit (ODD/SIGN): Select channel (0 or 1)
# Fourth bit (MSFB): 0 for LSB first
# Next 12 bits: 0 (don't care)
msg = 0b11
msg = ((msg<< 1) + adc_ch) << 5
msg = [msg, 0b00000000]
reply = spi.xfer2(msg)

    # Construct single integer out of the reply (2 bytes)
adc = 0
for n in reply:
adc = (adc<< 8) + n

    # Last bit (0) is not part of ADC value, shift to remove it
adc = adc>> 1

    # Calculate voltage form ADC value
voltage = (vref * adc) / 1024

return voltage

# Report the channel 0 and channel 1 voltages to the terminal
try:
while True:
    adc_0 = read_adc(0)
    adc_1 = read_adc(1)
print("Ch 0:", round(adc_0, 2), "V Ch 1:", round(adc_1, 2), "V")
time.sleep(0.2)

finally:
GPIO.cleanup()
```

Fig. 3.9 Python code snippet mcp3008

Figure 3.10 elucidates the circuit design of raspberry pi with the mcp3008 ic.

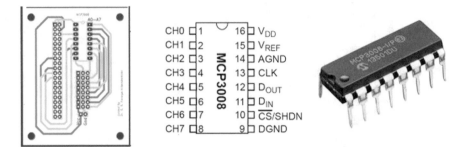

Fig. 3.10 RPI connection with MCP3008 IC

3.2   Localization and Mapping

The usage of robots and drones is exponentially increasing in day-to-day life for automation and sophistication purposes. The present state of the art in identification of these autonomous devices are based on Simultaneous Localization and Mapping(SLAM). Many similar algorithms have come; in outdoor-based mapping the lidar data is synchronized with the GPS data and computations are carried out for merging and developing Digital Elevation Models (DEM).

3.3   RP-Lidar

The following section mainly deals with simple RP-Lidar mainly used for position tracking and mapping purpose. This section mainly deals with the RPI lidar programming with rpi and mapping a simple two-dimensional area. RPLIDAR A1M8-R5 is a 2DLiDAR with 360° scanning. It was developed by SLAMTEC. It is relatively low cost. RPLIDAR A1M8-R5 can perform scans within 12-meter range. RPLIDAR A1M8-R5 produces 2D point cloud data that can be used in many applications such as mapping and localization. RPLIDAR A1M8-R5 can work well in all kinds of environments without direct sunlight exposure.

RPLIDAR A1M8-R5 is based on laser triangulation ranging principle. The device uses a high-speed vision acquisition and processing technology developed by SLAMTEC. The core of RPLIDAR A1M8-R5 runs clockwise to perform a 360-degree omnidirectional laser range scanning for its surrounding environment and then generate a map for that environment. Figure 3.11 provides the real-time picture and connection of RP-Lidar with RPi-4

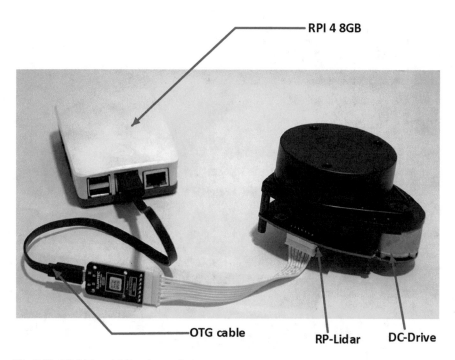

Fig. 3.11 RP-Lidar with Raspberry pi

Figure 3.12 elucidates the python script to activate the RP-Lidar

```python
#!/usr/bin/env python3
'''Animates distances and measurment quality'''
from rplidar import RPLidar
import matplotlib.pyplot as plt
import numpy as np
import matplotlib.animation as animation

PORT_NAME = '/dev/ttyUSB0'
DMAX = 4000
IMIN = 0
IMAX = 50

def update_line(num, iterator, line):
    scan = next(iterator)
    offsets = np.array([(np.radians(meas[1]), meas[2]) for meas in scan])
    line.set_offsets(offsets)
    intens = np.array([meas[0] for meas in scan])
    line.set_array(intens)
    return line,

def run():
    lidar = RPLidar(PORT_NAME)
    fig = plt.figure()
    ax = plt.subplot(111, projection='polar')
    line = ax.scatter([0, 0], [0, 0], s=5, c=[IMIN, IMAX],
                            cmap=plt.cm.Greys_r, lw=0)
    ax.set_rmax(DMAX)
    ax.grid(True)

    iterator = lidar.iter_scans()
    ani = animation.FuncAnimation(fig, update_line,
        fargs=(iterator, line), interval=50)
    plt.show()
    lidar.stop()
    lidar.disconnect()

if __name__ == '__main__':
    run()
```

Fig. 3.12 Code for running RPLiDAR and displaying output

Figure 3.13 illustrates the cloud points being scanned, straight line indicates the position of wall.

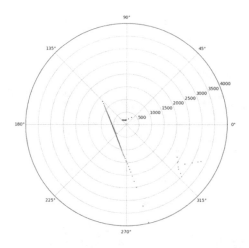

Fig. 3.13 Displaying RPLiDAR output

Figure 3.14 elucidates the code to measure the speed of the lidar.

```python
#!/usr/bin/env python3
'''Measures sensor scanning speed'''
from rplidar import RPLidar
import time

PORT_NAME = '/dev/ttyUSB0'

def run():
    '''Main function'''
    lidar = RPLidar(PORT_NAME)
    old_t = None
    data = []
    try:
        print('Press Ctrl+C to stop')
        for _ in lidar.iter_scans():
            now = time.time()
            if old_t is None:
                old_t = now
                continue
            delta = now - old_t
            print('%.2f Hz, %.2f RPM' % (1/delta, 60/delta))
            data.append(delta)
            old_t = now
    except KeyboardInterrupt:
        print('Stoping. Computing mean...')
        lidar.stop()
        lidar.disconnect()
        delta = sum(data)/len(data)
        print('Mean: %.2f Hz, %.2f RPM' % (1/delta, 60/delta))

if __name__ == '__main__':
    run()
```

Fig. 3.14 Code to measure RPLiDAR speed

Figure 3.15 elucidates the speed identified with the LIDAR interfaced with the raspberry pi 4 with Ubuntu OS.

Fig. 3.15 Display of consecutive measurements of RPLiDAR speed

References

1. V. Ziemann, *A hands-on course in sensors using the Arduino and Raspberry Pi* (2018). https://www.routledge.com/A-Hands-On-Course-in-Sensors-Using-the-Arduino-and-Raspberry-Pi/Ziemann/p/book/9780815393603
2. J. Lazar, *Arduino and LEGO projects* (Apress, New York, 2013)
3. E. Premeaux, B. Evans, *Arduino projects to save the world* (Apress, New York, NY, 2011)
4. P. Di Justo, E. Gertz, *Atmospheric monitoring with arduino: building simple devices to collect data about the environment*, 1st edn. (O'Reilly, Farnham, 2013)
5. C.A. Bell, *Beginning sensor networks with Arduino and Raspberry Pi.* (Apress, New York, 2013)
6. C. Bell, *Beginning Sensor Networks with XBee, Raspberry Pi, and Arduino Sensing the World with Python and MicroPython* (2020). https://www.apress.com/gp/book/9781484257951
7. K. Karvinen, T. Karvinen, *Getting started with sensors* (2014). https://www.oreilly.com/library/view/getting-started-with/9781449367077/#:~:text=To%20build%20electronic%20projects%20that,useful%20and%20educational%20sensor%20projects
8. J. Fraden, *Handbook of Modern Sensors: Physics, Designs, and Applications* (Springer, New York, NY, 2004)
9. J. Fraden, *Handbook of modern sensors: physics, designs, and applications*, 4th edn. (Springer, New York, 2010)
10. M. Margolis, *Make an Arduino-controlled robot*, 1st edn. (O'Reilly, Sebastopol, CA, 2013)

11. K. Karvinen, T. Karvinen, *Make Arduino bots and gadgets: learning by discovery. Sebastopol* (O'Reilly, CA, 2011)
12. H. Timmis, *Practical Arduino engineering* (Apress, Springer, New York, NY, 2011)
13. R. Anderson, D. Cervo, *Pro Arduino* (Apress, Berkeley, CA, 2013)
14. S. Monk, *Programming arduino next steps: going further with sketches* (2014). https://www.oreilly.com/library/view/programming-arduino-next/9781260143256/
15. M. Schwartz, O. Manickum, *Programming Arduino with Labview: build interactive and fun learning projects with* (2015). https://www.worldcat.org/title/programming-arduino-with-labview-build-interactive-and-fun-learning-projects-with-arduino-using-labview/oclc/903511516
16. P. Desai, *Python programming for Arduino develop practical Internet of things prototypes and applications with Arduino and Python* (Packt Publishing, Birmingham, UK, 2015)
17. A.K. Dennis, *Raspberry Pi Home Automation with Arduino* (Packt Publishing, Birmingham, 2015)
18. W.Y. Du, *Resistive* (CRC Press, Boca Raton, 2014)
19. D. Patranabis, *Sensors and transducers*, 2nd edn. (PHI Learning Private, Delhi, 2003)
20. S. Soloman, *Sensors handbook* (McGraw-Hill, New York, 2010)
21. *Wearable sensors fundamentals, implementation and applications* (Academic Press, San Diego, CA, 2014). https://www.elsevier.com/books/wearable-sensors/sazonov/978-0-12-819246-7
22. G. Guillen, *Sensor projects with raspberry PI: internet of things and digital image processing* (Apress, Berkeley, CA, 2020)
23. R. Gajjar, *Raspberry Pi Sensors: integrate sensors into your Raspberry Pi projects and let your powerful microcomputer interact with the physical world* (2015). https://dl.acm.org/doi/book/10.5555/2821177
24. T. Karvinen, K. Karvinen, V. Valtokari, *Make: sensors* (Maker Media, Sebastopol, CA, 2014)
25. S. Watkiss, *Learn electronics with Raspberry Pi: physical computing with circuits, sensors, outputs, and projects* (Apress, Springer Science+Business Media, New York, NY, 2016)

Chapter 4
Actuators Used in Rapid Prototyping

4.1 Introduction

Actuators are used in smart city applications like health care, smart agriculture, and industrial applications. Actuators are also automated in some cases to form Wireless Actuator Sensor Network (WASN). The tiny actuators are normally connected directly with the microcomputers; however actuators that are processing heavy loads are connected via driver circuits. The actuators are classified with the moving type as linear moving actuators and rotational actuators. These actuators satisfy many of our daily needs such as fan and escalators.

Based on the source of power, they are also classified as electric, pneumatic, and hydraulic. The most commonly used actuator is electric in nature; the actuators such as servo and stepper motors are controlled with the system generated pwm signals. Changes in width, time, and frequency of pulses can change the rotation speed, direction, and torque exerted by corresponding actuators [1–12]. Most of the actuators in industries are controlled with PLC-based circuits. This chapter mainly focuses on prototyping with very small servos and steppers which are normally done with arduino and raspberry pi boards. Figure 4.1 elucidates different types of actuators being discussed in this chapter, which includes micro stepper, servo sg90, infrared leds, visible light leds, and relay units. These actuators are mainly controlled with arduino and raspberry pi programming concepts for readers understanding [13–19].

© The Author(s), under exclusive license to Springer Nature Switzerland AG 2021
G. R. Kanagachidambaresan, *Role of Single Board Computers (SBCs) in rapid IoT Prototyping*, Internet of Things, https://doi.org/10.1007/978-3-030-72957-8_4

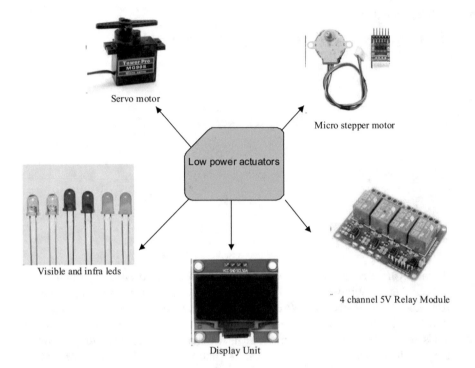

Fig. 4.1 Low-power actuators for IoT applications and conceptual designs

Table 4.1 provides the operation parameters and applications of different types of actuators

Table 4.1 Actuator type and their characteristics

S.no	Actuator type	Output	Operating range	Power required	Number of wire connection	External battery source required	Connectivity type Analog/i2c/spi, digital/ driver requirement etc.,
1	Servo motor	Angular motion	0V5 v	Less than 1 W	3	Not necessary for low power	PWM
2	Stepper motor	Angular position	2.55–2.8 V	1.68 A	4	Yes	Drivers or DPDT relay
3	Oled display	Digital display	3.3–5 V 20 mA max	3.3 V	12C pins	Yes	I2C and SPI

(continued)

Table 4.1 (continued)

S.no	Actuator type	Output	Operating range	Power required	Number of wire connection	External battery source required	Connectivity type Analog/i2c/spi, digital/ driver requirement etc.,
4	Relay	Mechanical contacts and solid state	5–20 ms	5 V	3	No (unless it is 12 V relay)	Both digital and analog
5	Solenoid	Mechanical energy angular motion	5–10 ms 15–150 ms	9–65 W	2 terminals each	External regulator or separate power supply	Both digital and analog
6	IR led	Light	760–1 nm Max = 780–50 nm	1.2–3.4 V	2 pins for each	No	Both digital and analog
7	Heat exchanger	Thermal	1000 °C and 1000 bars	Not needed	Wires not needed	Acid batteries	Tube box

| 8 | Speaker | Sound pressure level (SPL) in dB | 60 Hz to 18 kHz | 100–1000 W | 2 or more | No | Digital |
| 9 | Linear actuator | Straight line motion/ mechanical energy | 5–10,000 lb force | 15–2000 pounds | 4 pins | DC Battery | Digital |

4.2   Dc Motor

The operation of simple 5 V dc motor capable to operate less than 16 mA is discussed here. These tiny little motors are mainly used in remote operating cars, movable hands, and other low-torque applications. The program is capable to control the motor in forward and reverse directions. Here the direction of the motor is mainly controlled with armature-based control, where pin number 18 and 22 are connected with the armature. Code 1 DC motor forward and reverse programming in RPi is given below.

```
import RPi.GPIO asGPIO
import time

PIN_MOTOR1 = 18
PIN_MOTOR2 = 22

def forward ():
   GPIO.output (PIN_MOTOR1, GPIO.HIGH)
   GPIO.output (PIN_MOTOR2, GPIO.LOW)

def backward ():
   GPIO.output (PIN_MOTOR1, GPIO.LOW)
   GPIO.output (PIN_MOTOR2, GPIO.HIGH)

def stop ():
   GPIO.output (PIN_MOTOR1, GPIO.LOW)
   GPIO.output (PIN_MOTOR2, GPIO.LOW)

def setup ():
   GPIO.setmode (GPIO.BOARD)
   GPIO.setup (PIN_MOTOR1, GPIO.OUT)
   GPIO.setup (PIN_MOTOR2, GPIO.OUT)

print "starting"
set up()
while True:
   print "forward"
   forward ()
   time.sleep (2)
   print "backward"
   backward ()
   time.sleep (2)
   print "stop"
   Stop()
   time.sleep (2)
```

Code 1 Python Code snippet Motor actuation

The following section mainly discusses the arduino-based dc motor control. The program also includes braking concept of motor; furthermore the program can also be developed with P, PI, and PID controller for better controlling of motor. Code 2 provides arduino program for DC motor actuation.

```
constintpwm = 2 ;//initializing pin 2 as pwm
constint in_1 = 8 ;
constint in_2 = 9 ;

//For providing logic to L298 IC to choose the direction of the DC motor

void setup()
{
pinMode(pwm,OUTPUT) ;      //we have to set PWM pin as output
pinMode(in_1,OUTPUT) ;     //Logic pins are also set as output
pinMode(in_2,OUTPUT) ;
}

void loop()
{
//For Clock wise motion , in_1 = High , in_2 = Low

digitalWrite(in_1,HIGH) ;
digitalWrite(in_2,LOW) ;
analogWrite(pwm,255) ;

/*setting pwm of the motor to 255
we can change the speed of rotaion
bychaningpwm input but we are only
using arduino so we are using higest
value to driver the motor  */

//Clockwise for 3 secs
delay(3000) ;

//For brake
digitalWrite(in_1,HIGH) ;
digitalWrite(in_2,HIGH) ;
delay(1000) ;

//For Anti Clock-wise motion -IN_1 = LOW , IN_2 = HIGH
digitalWrite(in_1,LOW) ;
digitalWrite(in_2,HIGH) ;
delay(3000) ;

//For brake
digitalWrite(in_1,HIGH) ;
digitalWrite(in_2,HIGH) ;
delay(1000) ;
}
```

Code 2 Arduino C for DC motor driving

The following section mainly deals with the 4 bit and 8 bit lcd display programming using RPi and arduino. Figure 4.2 elucidates the connection diagram of RPi with light and contrast control.

4.3   LCD Programming with Raspberry Pi

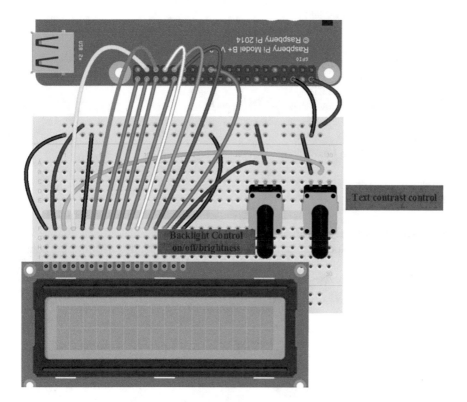

Fig. 4.2 Wiring diagram of RPi with LCD display.

The following table provides code for 4, 8 bit and text blinking in LCD with rpi. Code 3 provides RPi-LCD display python codes.

in 8 bitmode
from RPLCD import CharLCD
lcd = CharLCD(cols=16, rows=2, pin_rs=37, pin_e=35, pins_data=[40, 38, 36, 32, 33, 31, 29, 23])
lcd.write_string(u'Hello world!')

in 4 bitmode

from RPLCD import CharLCD
lcd = CharLCD(cols=16, rows=2, pin_rs=37, pin_e=35, pins_data=[33, 31, 29, 23])
lcd.write_string(u'Hello world!')

text blinking
import time
from RPLCD import CharLCD
lcd = CharLCD(cols=16, rows=2, pin_rs=37, pin_e=35, pins_data=[33, 31, 29, 23])

while True:
lcd.write_string(u"Hello world!")
time.sleep(1)
lcd.clear()
time.sleep(1)

Code 3 RPi-LCD display python codes

Figure 4.3 elucidates the connection diagram of LCD display with arduino board.

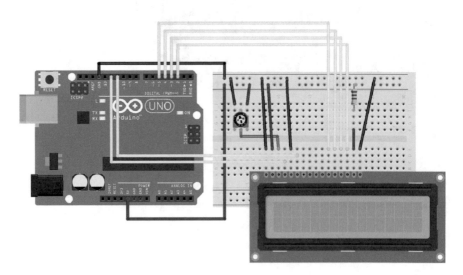

Fig. 4.3 Arduino connection with LCD display

Code 4 is for arduino to display hello world in LCD display.

The circuit:
* LCD RS pin to digital pin 12
* LCD Enable pin to digital pin 11
* LCD D4 pin to digital pin 5
* LCD D5 pin to digital pin 4
* LCD D6 pin to digital pin 3
* LCD D7 pin to digital pin 2
* LCD R/W pin to ground
* LCD VSS pin to ground
* LCD VCC pin to 5V
* 10K resistor:
* ends to +5V and ground
* wiper to LCD VO pin (pin 3)

```
// include the library code:
#include <LiquidCrystal.h>

// initialize the library by associating any needed LCD interface pin
// with the arduino pin number it is connected to
constintrs = 12, en = 11, d4 = 5, d5 = 4, d6 = 3, d7 = 2;
LiquidCrystallcd(rs, en, d4, d5, d6, d7);

void setup() {
  // set up the LCD's number of columns and rows:
lcd.begin(16, 2);
  // Print a message to the LCD.
lcd.print("hello, world!");
}

void loop() {
  // set the cursor to column 0, line 1
  // (note: line 1 is the second row, since counting begins with 0):
lcd.setCursor(0, 1);
  // print the number of seconds since reset:
lcd.print(millis() / 1000);
}
```

Code 4 Arduino LCD interface code

The following section deals with programming servo drive with RPi and arduino boards with Python and Arduino language. Figure 4.4 elucidates the connection diagram of rpi with servo motor. The orange wire in all servos is connected with PWM generator.

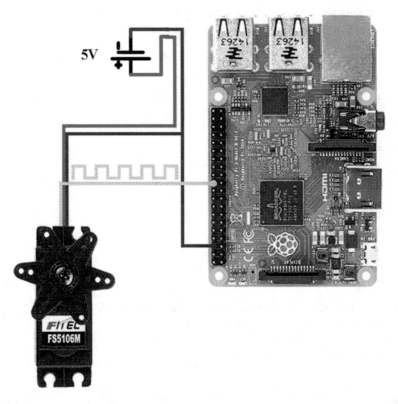

Fig. 4.4 Servo connection diagram

The program to drive the above circuit is briefly described in the following code 5.

```
importRPi.GPIO as GPIO
import time

P_SERVO = 22 # adapt to your wiring
fPWM = 50  # Hz (not higher with software PWM)
a = 10
b = 2

def setup():
globalpwm
GPIO.setmode(GPIO.BOARD)
GPIO.setup(P_SERVO, GPIO.OUT)
pwm = GPIO.PWM(P_SERVO, fPWM)
pwm.start(0)

defsetDirection(direction):
duty = a / 180 * direction + b
pwm.ChangeDutyCycle(duty)
print "direction =", direction, "-> duty =", duty
time.sleep(1) # allow to settle

print "starting"
setup()
for direction in range(0, 181, 10):
setDirection(direction)
direction = 0
setDirection(0)
GPIO.cleanup()
print "done"
```

Code 5 Code for PWM servo operation

Figure 4.5 provides the arduino-based servo connection diagram; here the pwn pin is connected with digital pin 10.

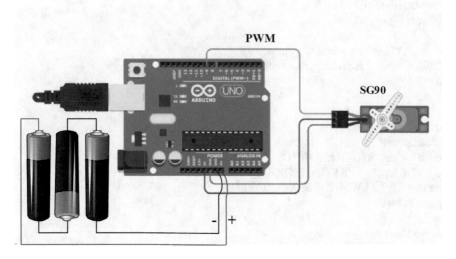

Fig. 4.5 Servo actuator circuit diagram

The arduino code 6 to drive the above circuit is discussed below.

```
// Include the servo library:
#include <Servo.h>
// Create a new servo object:
Servo myservo;
// Define the servo pin:
#define servoPin 9
// Create a variable to store the servo position:
int angle = 0;
void setup() {
// Attach the Servo variable to a pin:
myservo.attach(servoPin);
}
void loop() {
// Tell the servo to go to a particular angle:
myservo.write(90);
delay(1000);
myservo.write(180);
delay(1000);
myservo.write(0);
delay(1000);
// Sweep from 0 to 180 degrees:
for (angle = 0; angle <= 180; angle += 1) {
myservo.write(angle);
delay(15);
}
// And back from 180 to 0 degrees:
for (angle = 180; angle >= 0; angle -= 1) {
myservo.write(angle);
delay(30);
}
delay(1000);
}
```

Code 6 Arduino Code for servo operation

The following section deals with working with text to voice assistance using raspberry pi. This section is only done with raspberry pi 3 and above version of boards, because audio jack is not available with rpi zero and zero w boards. The text to speech option is achieved through pyttsx library in python. The driver alsa is also to be installed with apt-get command in order to proceed with the following section. Code 7 provides a very small code snippet to say "Role of Single Board computers" through audio jack with default female voice.

4.3.1   Python Text to Speech

```
import pyttsx3
engine = pyttsx3.init()
engine.say("Role of single board computers")
engine.runAndWait()
```

Code 7 Python code for text to speech
Following code 8 provides detailed information on speaking rate and also on tone of user, etc.

```
import pyttsx3
engine = pyttsx3.init() # object creation

""" RATE"""
rate = engine.getProperty('rate')   # getting details of current speaking rate
print (rate)                 #printing current voice rate
engine.setProperty('rate', 125)   # setting up new voice rate


"""VOLUME"""
volume = engine.getProperty('volume')   #getting to know current volume level (min=0 and max=1)
print (volume)                 #printing current volume level
engine.setProperty('volume',1.0)   # setting up volume level  between 0 and 1

"""VOICE"""
voices = engine.getProperty('voices')     #getting details of current voice
#engine.setProperty('voice', voices[0].id) #changing index, changes voices. o for male
engine.setProperty('voice', voices[1].id)  #changing index, changes voices. 1 for female

engine.say("Hello World!")
engine.say('My current speaking rate is ' + str(rate))
engine.runAndWait()
engine.stop()

"""Saving Voice to a file"""
# Onlinux make sure that 'espeak' and 'ffmpeg' are installed
engine.save_to_file('Hello World', 'test.mp3')
engine.runAndWait()
```

Code 8 Voice assistance in Python.

4.3.2   Self-Balanced Robot Design Using Sensors and Actuators

Figure 4.6 provides the circuit to create a two-wheel self-balanced robot.

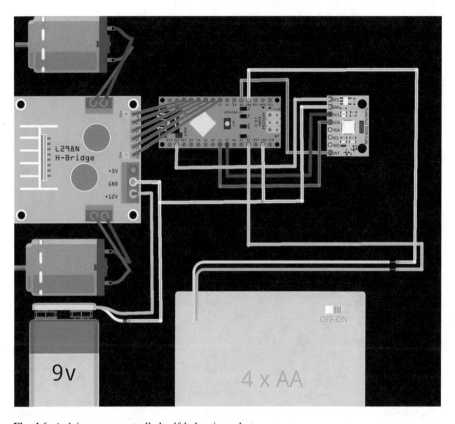

Fig. 4.6   Arduino nano controlled self-balancing robot

```
#include <PID_v1.h>
#include <LMotorController.h>
#include "I2Cdev.h"
#include "MPU6050_6Axis_MotionApps20.h"
#if I2CDEV_IMPLEMENTATION == I2CDEV_ARDUINO_WIRE
#include "Wire.h"
#endif
#define MIN_ABS_SPEED 20
MPU6050 mpu;
```

```
// MPU control/status vars
bool dmpReady = false; // set true if DMP init was successful
uint8_t mpuIntStatus; // holds actual interrupt status byte from MPU
uint8_t devStatus; // return status after each device operation (0 = success, !0 = error)
uint16_t packetSize; // expected DMP packet size (default is 42 bytes)
uint16_t fifoCount; // count of all bytes currently in FIFO
uint8_t fifoBuffer[64]; // FIFO storage buffer
Quaternion q; // [w, x, y, z] quaternion container
VectorFloat gravity; // [x, y, z] gravity vector
float ypr[3]; // [yaw, pitch, roll] yaw/pitch/roll container and gravity vector

//PID controller design
double originalSetpoint =173;
double setpoint = originalSetpoint;
double movingAngleOffset = 0.1;
double input, output;

//controller values kp,ki,kd, tune for user design
double Kp = 50;
double Kd = 1.4;
double Ki = 60;
PID pid(&input, &output, &setpoint, Kp, Ki, Kd, DIRECT);

double motorSpeedFactorLeft = 0.6;
double motorSpeedFactorRight = 0.5;
//drive  CONTROLLER
int ENA = 5;
int IN1 = 6;
int IN2 = 7;
int IN3 = 8;
int IN4 = 9;
int ENB = 10;
LMotorController motorController(ENA, IN1, IN2, ENB, IN3, IN4, motorSpeedFactorLeft,
motorSpeedFactorRight);
volatile bool mpuInterrupt = false; // indicates whether MPU interrupt pin has gone high
void dmpDataReady()
{
mpuInterrupt = true;
}

void setup()
{
Wire.begin();
TWBR = 24;// 400kHz I2C clock (200kHz if CPU is 8MHz)
```

```
Fastwire::setup(400, true);
#endif
mpu.initialize();
devStatus = mpu.dmpInitialize();
mpu.setXGyroOffset(220);
mpu.setYGyroOffset(76);
mpu.setZGyroOffset(-85);
mpu.setZAccelOffset(1788); // 1688 factory default for my test chip
if (devStatus == 0)
{
mpu.setDMPEnabled(true);
attachInterrupt(0, dmpDataReady, RISING);
mpuIntStatus = mpu.getIntStatus();
dmpReady = true;
packetSize = mpu.dmpGetFIFOPacketSize();


pid.SetMode(AUTOMATIC);
pid.SetSampleTime(10);
pid.SetOutputLimits(-255, 255);
}
else
{
Serial.print(F("DMP Initialization failed (code "));
Serial.print(devStatus);
Serial.println(F(")"));
}
}
```

```
void loop()
{
if (!dmpReady) return;
while (!mpuInterrupt && fifoCount < packetSize)
{
pid.Compute();
motorController.move(output, MIN_ABS_SPEED);
}
mpuInterrupt = false;
mpuIntStatus = mpu.getIntStatus();
fifoCount = mpu.getFIFOCount();
if ((mpuIntStatus & 0x10) || fifoCount == 1024)
{
mpu.resetFIFO();
Serial.println(F("FIFO overflow!"));
}
else if (mpuIntStatus & 0x02)
{
while (fifoCount < packetSize) fifoCount = mpu.getFIFOCount();
mpu.getFIFOBytes(fifoBuffer, packetSize);
fifoCount -= packetSize;
mpu.dmpGetQuaternion(&q, fifoBuffer);
mpu.dmpGetGravity(&gravity, &q);
mpu.dmpGetYawPitchRoll(ypr, &q, &gravity);
input = ypr[1] * 180/M_PI + 180;
}
}
```

Code 9 Self-stabilizing robot design using arduino nano

References

1. VipinaKumāra, Robot Operating System cookbook: over 70 recipes to help you master advanced ROS concepts (2018). https://www.perlego.com/book/771692/robot-operating-system-cookbook-over-70-recipes-to-help-you-master-advanced-ros-concepts-pdf
2. J.M. Zelle, *Python programming: an introduction to computer science* (Franklin, Beedle & Associates, Incorporated, Wilsonville, OR, 2004)
3. M. Hammond, A. Robinson, *Python programming on Win32*, 1st edn. (O'Reilly, Beijing, Sebastopol, CA, 2000)
4. K.D. Lee, *Python programming fundamentals* (Springer, London, New York, 2011)
5. P. Desai, *Python programming for Arduino develop practical Internet of things prototypes and applications with Arduino and Python* (Packt Publishing, Birmingham, UK, 2015)
6. M. Lutz, *Programming Python*, 2nd edn. (O'Reilly, Beijing, 2001)
7. G.S.V.A. *Introduction to Python programming*. (2019). https://www.routledge.com/Introduction-to-Python-Programming/S-A/p/book/9780815394372

8. J. Guttag, *Introduction to computation and programming using Python: with application to understanding data*, 2nd edn. (The MIT Press, Cambridge, Massachusetts, 2016)
9. M. Banzi, *Getting Started with Arduino*, 2nd edn. (O'Reilly & Associates, Sebastopol, CA, 2011)
10. J. Bayle, *C programming for Arduino learn how to program and use Arduino boards with a series of engaging examples, illustrating each core concept* (Packt Pub., Birmingham, 2013)
11. R. Singh, A. Gehlot, B. Singh, S. Choudhury, *Arduino Meets MATLAB: interfacing, programs and simulink* (Bentham Science Publishers, Sharjah, 2018)
12. F. Perea, *Arduino essentials: enter the world of Arduino and its peripherals and start creating interesting projects* (2015). https://www.oreilly.com/library/view/arduino-essentials/9781784398569/
13. B.R. Kent, *Science and computing with Raspberry Pi* (2018). https://iopscience.iop.org/book/978-1-6817-4996-9.pdf
14. S. Monk, I. OverDrive, *Raspberry Pi Cookbook* (O'Reilly Media, Sebastopol, CA, 2016)
15. S. Monk, *Programming the Raspberry Pi: getting started with Python* (2013). https://www.accessengineeringlibrary.com/content/book/9781259587405
16. C.A. Philbin, *Adventures in Raspberry Pi®*, 2nd edn. (Wiley, Chichester, West Sussex, 2015)
17. S. McManus, M. Cook, *Raspberry Pi for dummies* (John Wiley & Sons, Hoboken, NJ, 2013)
18. N.S. Yamanoor, S. Yamanoor, "High quality, low cost education with the Raspberry Pi," in *2017 IEEE Global Humanitarian Technology Conference (GHTC)*, San Jose, CA (Oct. 2017). pp. 1–5. https://doi.org/10.1109/GHTC.2017.8239274
19. E. Upton, J. Duntemann, R. Roberts, T. Mamtora, B. Everard, *Learning computer architecture with Raspberry Pi* (Wiley, Indianapolis, IN, 2016)

Chapter 5
Introduction to Wired and Wireless IoT Protocols in SBC

5.1 Introduction

There are lots of wired and wireless protocols especially facilitating Internet of Things environments. When the nodes inside the environment are immobile, high data communication requirement and handling trusted data, wired networks is the best case. Most of the Industrial Internet of Things (IIoT) systems provide wired communication for facilitating their needs. Figure 5.1 illustrates the network classification mainly used in constructing rapid iot using SBCs [1, 2].

© The Author(s), under exclusive license to Springer Nature Switzerland AG 2021
G. R. Kanagachidambaresan, *Role of Single Board Computers (SBCs) in rapid IoT Prototyping*, Internet of Things, https://doi.org/10.1007/978-3-030-72957-8_5

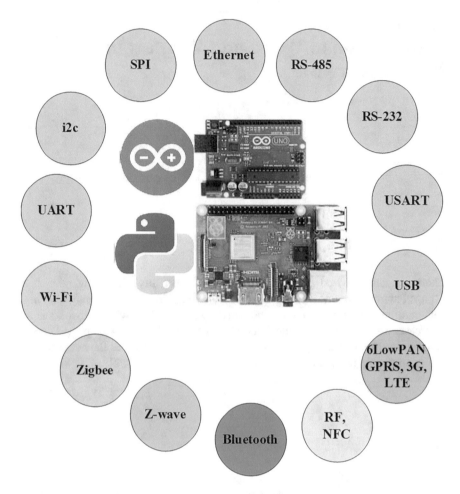

Fig. 5.1 Wired and Wireless routing protocols for Single Board Computers

The most commonly used wired internal protocols in prototyping includes (a) i2c, (b) SPI and, (c) UART [3–6]

5.2 I2C

Inter Integrated Circuit bus is mainly designed for real-time clocks control interface and has three different speed, 100 kbps, Fast 400 kbps, and High speed 3.4 Mbps. It is a wire protocol, Vcc, Gnd, SDA (Serial Data), and SCL (Serial Clock). The i2c operated with 5 V can be connected with low number of wire connectivity. The data communication will be amicable when the i2c devices operate occasionally. The scheme used for addressing allows multiple devices to be connected. The system however cannot be used for very large scale devices control and operation. Figure 5.2 provides the packet information details.

Start Bit	Device Slave address	Ack	Internal register	Ack	Data	Stop

Fig. 5.2 Packet information details of i2c

```
//MASTER
#include <Wire.h>
void setup() {
Wire.begin();        // join i2c bus (address optional for
master)
Serial.begin(9600); // Initialize data communication in
9600 baud rate
}
void loop() {
Wire.requestFrom(8, 6);   // request 6 bytes from slave
device #8
while (Wire.available()) { //checking data availability
char c = Wire.read(); // receive data
Serial.print(c);        // display the data
 }
delay(500);
}
```

Code 1 Master communication code

```
//SLAVE
#include <Wire.h> //i2c library
void setup() {
Wire.begin(8);            // joining the network with 8bit
address
Wire.onRequest(requestEvent); // event registration
}
void loop() {
delay(100);
}
voidrequestEvent() {
Wire.write("Single Board Computers"); // data response
}
```

Code 2 Slave communication code.

Figures 5.3 and 5.4 provide circuit connection and PCB design of I2C communication with arduino.

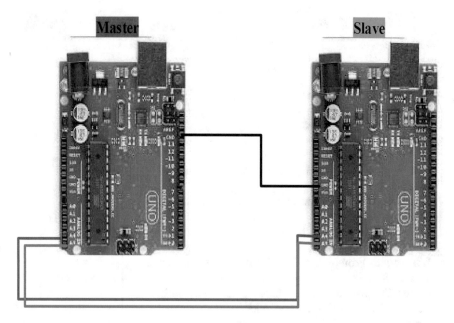

Fig. 5.3 I2C communication sharing data between two arduinos

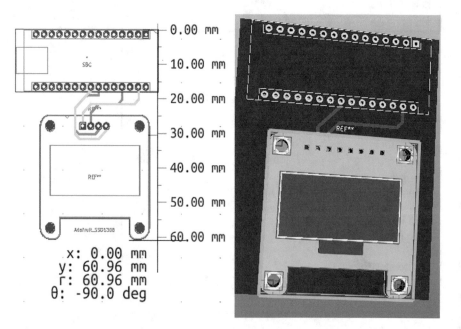

Fig. 5.4 Arduino Connection with OLED display Kicad design

The i2c connection in raspberry pi is connected and can be programmed with the code 3 given below.

Figure 5.5 provides the kicad design of RPi interface with adc1115 i2c IC.

```
from time import sleep
import Adafruit_ADS1x15
adc = Adafruit_ADS1x15.ADS1115()
GAIN = 1
print('Reading ADS1x15 values, press Ctrl-C to quit...')
print('| {0:>6} | {1:>6} | {2:>6} | {3:>6} |'.format(*range(4)))
print('-' * 37)
while True:
values = [0]*4
for i in range(4):
    # Read the specified ADC channel using the previously set gain value.
values[i] = adc.read_adc(i, gain=GAIN)
  print('| {0:>3} | {1:>3} | {2:>3} | {3:>3} |'.format(*values))
sleep(1)
```

Code 3 ADS1115 rpi interface code
3d model Circuit Design

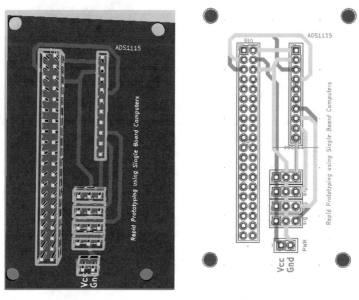

3d model   **Circuit Design**

Fig. 5.5 Kicad circuit of i2c

5.3   Serial Peripheral Interface (SPI) control for Arduino and Raspberry pi

The Serial Peripheral Interface (SPI) works on synchronous serial communication mainly used for short distance. The Serial Clock, Master output slave input, master input slave output, and slave select pins mainly used to connect slave devices. The SPI is mainly faster in case of asynchronous serial communication; however it requires more physical components. The slaves are unable to share the data; it can only report data to the master node. Figure 5.6 elucidates the simple SPI communication architecture. This section mainly focuses on SPI communication on arduino and raspberry pi based single board computer.

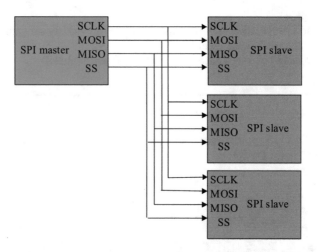

Fig. 5.6 SPI connection with Master and multiple slave

A simple design to interface 8 sensor data using MCP3008 8 channel 10bit adc converter is demonstrated in Fig. 5.7. The code 4 provides the python code for MCP3008 interface.

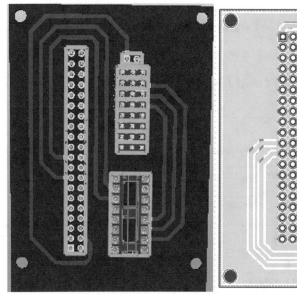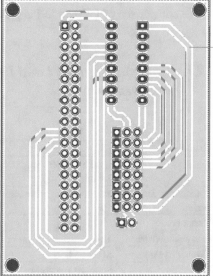

Fig. 5.7   Kicad circuit design of SPI communication

```
from time import sleep
importAdafruit_GPIO.SPI as SPI
import Adafruit_MCP3008
CLK  = 18
MISO = 23
MOSI = 24
CS   = 25
mcp = Adafruit_MCP3008.MCP3008(clk=CLK, cs=CS, mi-
so=MISO, mosi=MOSI)
print('Reading MCP3008 values, press Ctrl-C to quit...')
print('| {0:>4} | {1:>4} | {2:>4} | {3:>4} | {4:>4} | {5:>4} |
{6:>4} | {7:>4} |'.format(*range(8)))
print('-' * 57)
while True:
values = [0]*8
for i in range(8):
values[i] = mcp.read_adc(i)
    # Print the ADC values.
    print('| {0:>4} | {1:>4} | {2:>4} | {3:>4} | {4:>4} | {5:>4} |
{6:>4} | {7:>4} |'.format(*values))
    # Pause for half a second.
sleep(0.5)
```

Code 4 ADCMCP3008 Python RPi Interface code

The comparison of I2C and SPI is given as follows.

5.3.1 Comparison of SPI and I2C Communication Merits and Demerits

Characteristics	I2C	SPI
Speed	100 kbps, 400 kbps, 1mbps, 3.4 mbps	1mbps, 10 mbps can be reached to100 mbps
Communication mode	half duplex/synchronous	Full/synchronous
Master config	Multi master	Single
Data acknowledge	Yes	No
Number of Pins	SDA, SCL	MISO, MOSI, CLK, and CS
Scalability	Supports scalability	Does not support scalability
Overhead data in wire	More ack details and overhead	Less overhead details
Noise sensitivity	High	Low

5.4 UART Data Communication with NXP Nucleo 32 Board and Raspberry pi

The following section provides UART data communication. Figure 5.8 provides the kicad design circuit with I2C and UART.

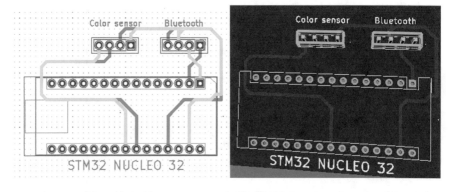

Fig. 5.8 Kicad Circuit design for i2c and uart combination

This circuit can be used to sense data from any i2c sensor and report the data via HC-05 bluetooth device.

Mbed c program for Nucleo-32

```
#include "mbed.h"
Serial pc(SERIAL_TX, SERIAL_RX);
I2C i2c(I2C_SDA, I2C_SCL);
AnalogInain(A0);
int main() {
while(1) {
wait(2);
float f=ain.read();
pc.printf("value is: %f \n",f);
i2c.write(f);
  }
}
```

Code 5 Mbed C program for Nucleo 32

5.5   UART Rpi Communication and Programming

Some of the sensors and devices are connected with UART communication; the pin connections are Vcc, Gnd, Tx, and Rx. Devices like Bluetooth, Pen drive, and other single board to single board communication can be done through UART communication. The devices are connected with cross communication rx of device A to Tx of Device B and vice versa for communication. Figure 5.9 illustrates the kicad circuit design and 3d view of circuit design.

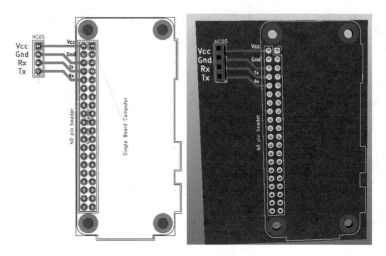

Fig. 5.9 UART kicad circuit design with Rpi

Code 6 Serial data communication Rx, Tx python code

```
import serial
from time import sleep
ser = serial.Serial ("/dev/ttyS0", 9600)    //port s0 is opened with 9600 baud rate
while True:
received_data = ser.read()          #data read from serial port
sleep(0.03)
data_left = ser.inWaiting()          #remaining byte check
received_data += ser.read(data_left)
print (received_data)                 #Display received data
ser.write(received_data)              #serial data transmit
```

Table 5.1 illustrates the comparison of three universal data protocols.

Table 5.1 UART, USART, and USB

Mode of communication	UART	USART	USB
Expansion	Universal Asynchronous transmitter and receiver	Universal Synchronous and asynchronous data transmitter and receiver	Universal serial Bus
Number of Wire	Tx, Rx	Rx and Tx	D+, D−
Overhead	Along with Pulses	Along with Class pulse	with Clock pulse
Mode	Half duplex	Full duplex	Full duplex
Speed	Slow than USART	Slow than USB	Fast than others

5.6   RS485 Protocol

RS485 protocol is mainly used for data transfer for up to 1 km or 4000 ft communication. The rs485 methodology is low cost and has high immune to signal noises. It operates with two signals Tx and Rx, and operates mainly on half duplex master slave mode operation. Figure 5.10 illustrates the simple arduino to arduino data communication within 1km distance with UART to RS485 converter. The converter mainly aids industrial applications and for Industry 4.0 standards.

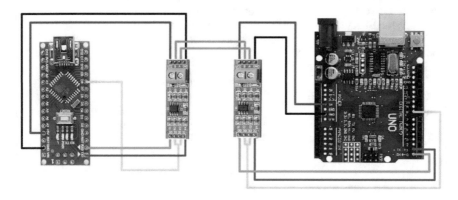

Fig. 5.10 Master slave data communication in arduino pair

Code 7 Master Sender Code at Arduino Nano

```
#define LED      13   // LED bulp declaration
#define MASTER_EN  8  // Enable pin to connect rs485
void setup() {
pinMode(LED , OUTPUT);          // LED pin declared as output
pinMode(MASTER_EN , OUTPUT);    // Declare Enable of RS485 pin
as output
Serial.begin(9600);             // baudrate for data communication
digitalWrite(MASTER_EN , LOW);  // Low the enable pin
                 // Mode on receive pin
}
void loop() {
digitalWrite(MASTER_EN , HIGH);   // Enable pin high to send Data
delay(5);                 // delay of 5ms minimum is set
Serial.println('A');           // Send A- char
Serial.flush();              // wait for transmission of data
delay(1000);
digitalWrite(MASTER_EN , LOW);    // Receiving-mode ON
}
```

Code 8 Receiver code at Arduino Uno
Slave code

```
#define LED       13
#define SLAVE_EN  8

void setup() {
pinMode(LED , OUTPUT);                  // LED pin - output
pinMode(SLAVE_EN , OUTPUT);                // Enable pin (output)
Serial.begin(9600);                    // baud rate for serial communication
digitalWrite(SLAVE_EN , LOW);            // Enable Pin low
                        // Receiving mode-ON
}
void loop() {
while(Serial.available())              // Enter the while loop if serial data available
  {
if(Serial.read() == 'A')            // A is available
  {
digitalWrite(LED , !digitalRead(LED));   // LED Blink
  }
  }
}
```

5.7   HC05 AT Command Setting

This section discusses about Bluetooth serial communication. Figure 5.11 elucidates arduino Bluetooth configuration.

Code 9 Bluetooth serial data communication

Arduino Program

```
#include <SoftwareSerial.h>
SoftwareSerialBTSerial(10, 11);
void setup()
{
pinMode(9, OUTPUT);
digitalWrite(9, HIGH);
Serial.begin(9600);
Serial.println("Enter AT commands:");
BTSerial.begin(38400);
}
void loop()
{
  if (BTSerial.available())
Serial.write(BTSerial.read());
  if (Serial.available())
BTSerial.write(Serial.read());
}
```

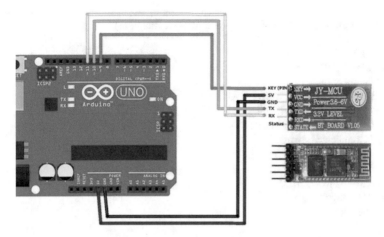

Fig. 5.11 HC-05 AT command connection with Arduino Uno

For configuring Bluetooth through rpi, the following command is executed in terminal window. The command will install picocom package in the RPi.

sudo apt-get install picocom.

The HC-05 is connected as per the connection diagram given in Fig. 5.12. The following command is executed in terminal to enter inside the minicom window.

picocom /dev/ttyAMA0 -b 38400 -i --omapcrcrlf–c

5.8 Bluetooth Interface and Communication with SBC

The Bluetooth protocol is utilized for transmitting and receiving data within 10 m distance. The Bluetooth data communications consist of scan pair and transmit data in simple. Normally used Bluetooth data communication device is HC-05; the circuit and explanation of HC-05 in arduino and Nucleo-32 boards were already discussed. The SBCs Raspberry pi includes indigenous Bluetooth low energy (BLE). Code 10 illustrates the python program for Bluetooth low energy protocol.

```
frombtpycom import *  # Standard-Python (PC, Raspi, ...)
defonStateChanged(state, msg):
global reply
if state == "CONNECTING":
print "Connecting", msg
elif state == "CONNECTION_FAILED":
print "Connection failed", msg
elif state == "CONNECTED":
print "Connected", msg
elif state == "DISCONNECTED":
print "Disconnected", msg
elif state == "MESSAGE":
print "Message", msg
reply = msg

serviceName = "EchoServer"
print "Performing search for service name", serviceName
client = BTClient(stateChanged = onStateChanged)
serverInfo = client.findService(serviceName, 20)
ifserverInfo == None:
print "Service search failed"
else:
print "Got server info", serverInfo
ifclient.connect(serverInfo, 20):
for n in range(0, 101):
client.sendMessage(str(n))
        reply = ""
while reply == "":
time.sleep(0.001)
client.disconnect()
```

Code 10 BLE-client python program.

5.9   Zigbee Communication

Zigbee communication is used to transmit low data rate sensor data to long range like 1–5 km. The new zigbee radios available in market are capable to transmit data even to 10 km with line of sight. The most commonly used zigbee radios in market belong to xbee, from DIGI company.

The following section mainly describes the configuration and mode of working of xbee radios with AT and API modes. Here, the AT mode is transparent mode capable to transmit live data to the other side. In case of API mode, i.e., Application Peripheral Interface Mode, the radios are capable to transmit data and commands.

The commercially available xbee radios include with different power capable to transmit data to long ranges; xbee, Xbee pro, and Xbee 3 options are available in the market (Table 5.2).

Table 5.2 Xbee RF radios

S. No	Xbee radio	Xbee radio	Range
1		XBee 3G	Cellular and LPWAN
2		XBee 3 LTE-M / NB-IoT	
3		XBee 3 LTE Cat-1	
4		XBee 868LP/900HP	Long range radios
5		XBee SX 900 / 868	
6		Xbee s2c	Short range (900/868 MHz/2.4 GHz)
7		XBEE3	

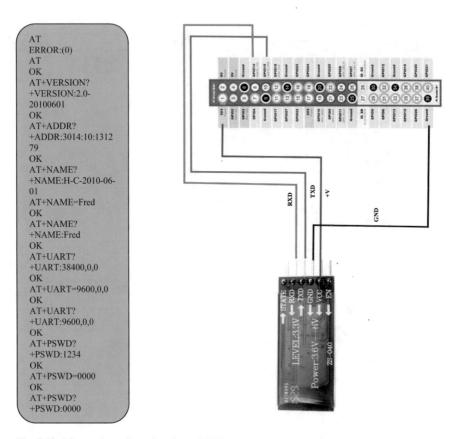

```
AT
ERROR:(0)
AT
OK
AT+VERSION?
+VERSION:2.0-
20100601
OK
AT+ADDR?
+ADDR:3014:10:1312
79
OK
AT+NAME?
+NAME:H-C-2010-06-
01
AT+NAME=Fred
OK
AT+NAME?
+NAME:Fred
OK
AT+UART?
+UART:38400,0,0
OK
AT+UART=9600,0,0
OK
AT+UART?
+UART:9600,0,0
OK
AT+PSWD?
+PSWD:1234
OK
AT+PSWD=0000
OK
AT+PSWD?
+PSWD:0000
```

Fig. 5.12 Bluetooth configuration through RPi

5.9.1   Configuring XBee Radios Using XCTU Software

The XCTU software can be downloaded from the link https://www.digi.com/prod-ucts/xbee-rf-solutions/xctu-software/xctu#productsupport-utilities.   The   XBee radios can be configured in three modes.

1. Application Transparent (AT) operating mode.
2. API operating mode.
3. API escaped operating mode.

Figure 5.13 elucidates the GUI for XCTU with palettes

Fig. 5.13   XCTU-SOFTWARE

The XBee on connecting with the USB adapter with the system will be recognized as a communication port. The symbol E indicates End device, R indicates router, and C indicates Coordinator. The coordinator has the capability to talk with all radios and broadcast the message. The MAC address of the radios is unique and is used for forwarding packet to corresponding ZigBEEradio. Figure 5.14 provides the Rpi interface with XBee.

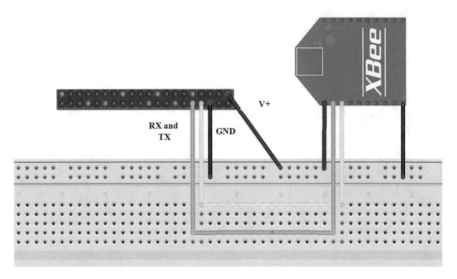

Fig. 5.14 XBEE connection with RPi

Raspberry Pi

import serial, time, datetime, sys
```
from xbee import XBee
ser = serial.Serial("/dev/ttyAMA0", 9600)
xbee = XBee(ser)
print 'Starting Up Tempature Monitor'
while True:
    try:
        response = xbee.wait_read_frame()
        print response
    except KeyboardInterrupt:
        break
ser.close()
```

Code 11 python code for xbeecomm

The above program verifies and displays the frame of the received packet. For sending strings and commands, the program on HC-05 can be used. For remote command transfer and enabling Xbee pin, the following chart is followed and frames are constructed based on the need. Figure 5.15 illustrates the frame generation and remote pin operation of Xbee using XCTU software. If the address is given as 00, the end device sends data only to sink and does not forward to other devices. If the address is given as FFFF, the message is broadcasted to all nodes within the PAN id.

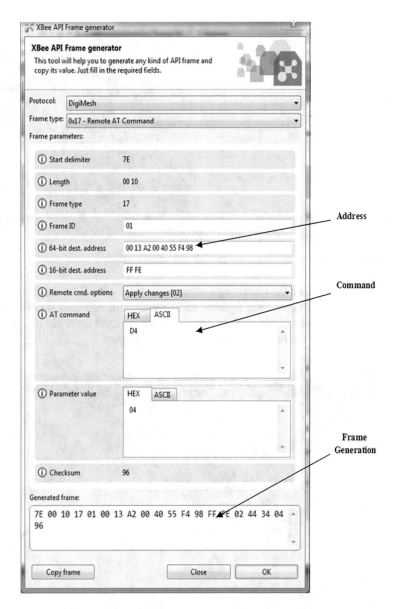

Fig. 5.15 Frame generation using XCTU

Figure 5.16 shows the kicad design circuit of the proposed xbee-based relay board

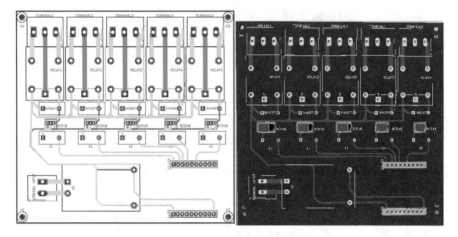

Fig. 5.16 Kicad design of the xbee-based remote relay control board.

Figure 5.17 provides the control option carried out in controlling the relay board with xbee communication. Here a user interface with python tkinter is designed to control the entire 5 relay. The 5 relays are connected with the timer circuit, so individual relays can be controlled with time package in python.

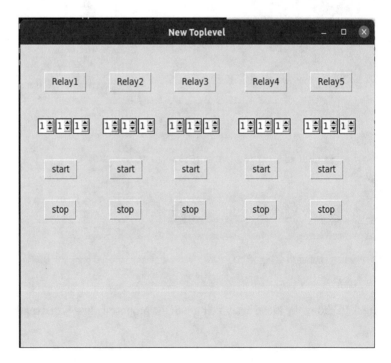

Fig. 5.17 Python-based GUI design for relay activation using timer control

Figure 5.18 provides the control architecture; the entire module can be controlled within 1 km radius; it can also be extended by replacing certain radios like xbee pro and xbee 3 modules.

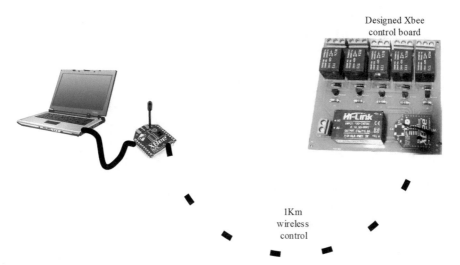

Fig. 5.18 Industry ready board controlled with xbee-based relay control

5.10 Conclusion

The various IoT wired and wireless networking programming modules were deeply discussed in this chapter. The important protocols used in industrial applications as well as smart city applications include rs485, Bluetooth, etc. The chapter also discusses on industry ready xbee controlled boards. The xbee control and data transfer with API and AT commands were also discussed in this chapter.

References

1. D.-S. Kim, H. Tran-Dang, *Industrial sensors and controls in communication networks: from wired technologies to cloud computing and the internet of things*, 1st edn. (Springer International Publishing : Imprint: Springer, Cham, 2019)
2. *International Conference on Information Networking and I. Chong, Information networking: international conference, ICOIN 2002*, Cheju Island, Korea, January 30-February 1, 2002 : revised papers (Berlin; Hong Kong, Springer-Verlag, 2002)
3. A. Bensky, *Short-range wireless communication fundamentals of RF system design and application* (Elsevier Science & Technology Books, Burlington, MA)

4. T. Cooklev, *Wireless communication standards: a study of IEEE 802.11, 802.15, and 802.16* (Standards Information Network, IEEE Press, New York, 2004)
5. T.L. Singal, *Wireless communications* (Tata Mcraw Hill Education Private Ltd., New Delhi, 2010)
6. Wireless communications theory and techniques (2004)

Chapter 6
Node-Red Programming and Page GUI Builder for Industry 4.0 Dashboard Design

6.1 Introduction

Node-red programming was initially developed by the JS foundation company for programming edge computing boards. Node-red is mainly designed as browser-based editor; this makes easy wiring of wide range of nodes. The node-red uses build in libraries to save multiple function assignments and templates that can be reused. The entire flow editing concept is designed with the node.js concept which is a lightweight runtime environment [1–5]. This flow editor can run on most of the edge devices like raspberry pi and other simple cloud architectures.

The node-red can be easily installed with the bash command or with the apt-get command. Entering the following command in the terminal window of raspberry pi with internet connection would install the node-red easily.

Command:

```
bash<(curl -sL https://raw.githubusercontent.com/node-red/linux-
installers/master/deb/update-nodejs-and-nodered)
```

Figure 6.1 illustrates the installation procedure of node-red in the raspberry pi terminal window. Once the command is entered, the installation procedure starts as shown in Fig. 6.2.

© The Author(s), under exclusive license to Springer Nature Switzerland AG 2021   121
G. R. Kanagachidambaresan, *Role of Single Board Computers (SBCs) in rapid IoT Prototyping*, Internet of Things, https://doi.org/10.1007/978-3-030-72957-8_6

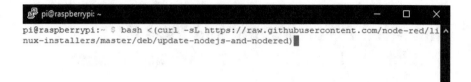

Fig. 6.1 Node-red installation bash script

Figure 6.2 elucidates the installation updates in terminal window.

Fig. 6.2 Node-red installation steps in terminal window

The following command is used to operate the node-red, the flow editor can be started and stopped with the start and stop operation.

```
node-red start
//node-red flow editor start option

node-red stop
//node-red flow editor stop option

node-red restart
//node-red restart during deadlock conditions
```

```
node-red log
//node-red log generation on various operations

sudosystemctl enable nodered.service
//automatic boot start of node-red

sudosystemctl disable nodered.service
//disabling auto boot start of node-red.
```

Once node-red service is started, the terminal window will start showing the log and response of each service in the terminal window. The flow editor can be viewed with the ip address of the raspberry pi with 1880 port number. The flow editor is mainly used for the wiring the terminals and programming with the functions. The output response of the flow editor can be viewed with the user interface window.

```
192.168.0.***:1880
//flow editor address.

192.168.0.***:1880/ui
//user interface dashboard address.
///*** to be replaced with Raspberry pi ip address.
```

Figure 6.3 elucidates the response of the node-red start command in the terminal window. Once the node-red is started, the flow editor can also be viewed with the local port address 127.0.0.1.:1880.

```
pi@raspberrypi: ~                                              —   □   ×
pi@raspberrypi:~ $ node-red start
23 Dec 15:51:13 - [info]

Welcome to Node-RED
===================

23 Dec 15:51:13 - [info] Node-RED version: v1.2.6
23 Dec 15:51:13 - [info] Node.js  version: v12.20.0
23 Dec 15:51:13 - [info] Linux 4.19.66+ arm LE
23 Dec 15:51:17 - [info] Loading palette nodes
23 Dec 15:51:29 - [info] Dashboard version 2.16.2 started at /ui
23 Dec 15:51:30 - [warn] rpi-gpio : Raspberry Pi specific node set inactive
23 Dec 15:51:30 - [warn] Missing node modules:
23 Dec 15:51:30 - [warn]  - node-red-contrib-ibm-watson-iot (0.2.8): wiotp-crede
ntials, wiotp in, wiotp out
23 Dec 15:51:30 - [warn]  - node-red-node-email (1.6.2): e-mail, e-mail in
23 Dec 15:51:30 - [warn]  - node-red-node-sentiment (0.1.3): sentiment
23 Dec 15:51:30 - [warn]  - node-red-node-twitter (1.1.5): twitter-credentials,
twitter in, twitter out
23 Dec 15:51:30 - [warn]  - node-red-node-feedparser (0.1.14): feedparse
23 Dec 15:51:30 - [warn]  - node-red-node-pi-sense-hat (0.0.18): rpi-sensehat in
, rpi-sensehat out
23 Dec 15:51:30 - [info] Removing modules from config
23 Dec 15:51:31 - [info] Settings file  : /home/pi/.node-red/settings.js
```

Fig. 6.3   node-red start updated in terminal window

Figure 6.4 elucidates the server configured and update shown in the terminal window of the raspberry pi.

```
pi@raspberrypi: ~                                                 —   □   ×
23 Dec 15:51:31 - [info] Settings file   : /home/pi/.node-red/settings.js
23 Dec 15:51:31 - [info] Context store   : 'default' [module=memory]
23 Dec 15:51:31 - [info] User directory : /home/pi/.node-red
23 Dec 15:51:31 - [warn] Projects disabled : editorTheme.projects.enabled=false
23 Dec 15:51:31 - [info] Flows file      : /home/pi/.node-red/start
23 Dec 15:51:31 - [info] Creating new flow file
23 Dec 15:51:31 - [warn]


---------------------------------------------------------------------

Your flow credentials file is encrypted using a system-generated key.

If the system-generated key is lost for any reason, your credentials
file will not be recoverable, you will have to delete it and re-enter
your credentials.

You should set your own key using the 'credentialSecret' option in
your settings file. Node-RED will then re-encrypt your credentials
file using your chosen key the next time you deploy a change.
---------------------------------------------------------------------

23 Dec 15:51:31 - [info] Starting flows
23 Dec 15:51:31 - [info] Started flows
23 Dec 15:51:31 - [info] Server now running at http://127.0.0.1:1880/
```

Fig. 6.4   Server configured in port 127.0.0.1:1880

Figure 6.5 provides the node-red console window.

Node-red
console start

Fig. 6.5   Rpi Node-red console screen 1

In order to design new packages in the node-red dashboard, the manage palette option is used in the node-red right hand side window. Let us install node-red dashboard in the raspberry pi; Fig. 6.6 illustrates the node-red dashboard palette being installed in the raspberry pi.

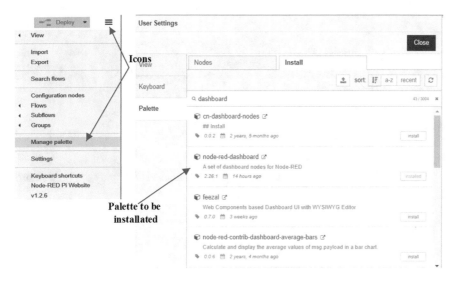

Fig. 6.6 Flow editor screen 1

A simple date time dashboard display is designed in the following section. Figure 6.7 illustrates the flow editor programmed with the components like

Fig. 6.7 Flow editor palettes

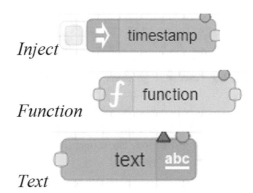

Figure 6.8 illustrates the time stamp, function, and text wired in node-red environment.

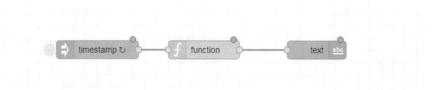

Fig. 6.8 Drag and Drop wired with text

Figure 6.9 elucidates the node java script function used to display the time stamp.

⚙ **Properties**			⚙ 📄 🖾

🏷 Name	Name		▣▾

Setup	**Function**	Close

```
1   msg.payload = new Date().toLocaleTimeString()
2   return msg;
```

Fig. 6.9 Function programming for time stamp conversion

Figure 6.10 elucidates the time stamp inject configuration, the time stamp is injected every second. The repeat operation is done in infinity loop with while loop.

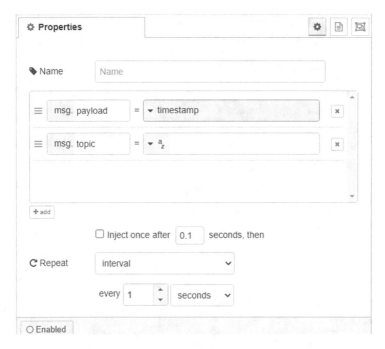

Fig. 6.10   Time stamp injection with repeated interval

Once all the points are configured, the text icon has to be configured with the name and UI icons. Figure 6.11 elucidates the icon configuration of text user interface.

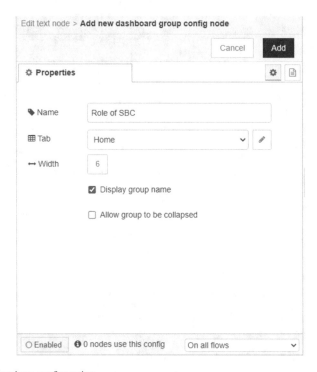

Fig. 6.11   Text icon configuration

Figure 6.12 elucidates the user interface designed and the clock details can be clearly viewed at 192.168.0.xxx:1880/ui.

Fig. 6.12   Time stamp and user interface designed with flow editor

Function to generate random number within 0–100 is given here.

```
msg.payload = Math.round(Math.random()*100);
returnmsg;
```

Figure 6.13 elucidates the simple dashboard being created with the flow editor to show various gauges that can be used for monitoring purpose.

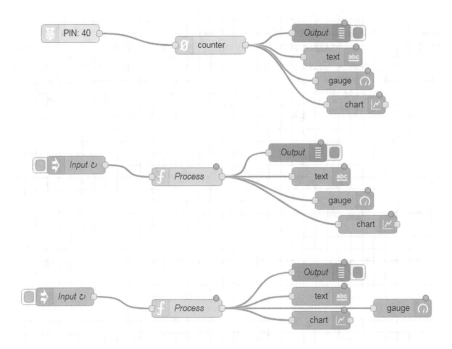

Fig. 6.13 Simple Dashboard design using flow editor

Here the process is programmed with the random math function as presented above to generate random numbers. The dashboard in the ui terminal is given in Fig. 6.14.

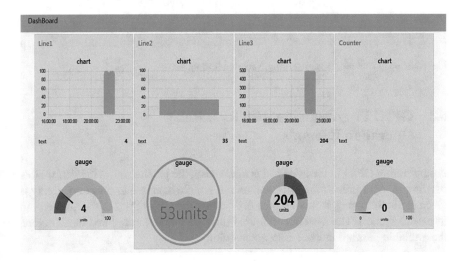

Fig. 6.14 Dashboard design for Industry 4.0 monitoring

6.2　Camera Image Acquisition with Rapsberry Pi

Apart from the dashboard designs and other sensor integration, raspberry pi-based cctv is designed in this following section. The raspicam camera is used to take pictures and monitor the same in the dashboard. Figure 6.15 elucidates the flow editor program for configuring raspberry pi camera with the switch command. Here the manual photo palette is used to take pictures every 3 sec, and it triggers the function to take pictures using raspicam. Once the shot is taken, the image is arranged for display purposed with corresponding width and height option.

Fig. 6.15　Flow editor program for raspicam activation

Figure 6.16 elucidates the user interface diagram of the raspicam cctv programming.

Fig. 6.16　Raspicam programming using Node-red flow editor

6.3　GPIO Programming with RPI: Relay Operation Example

Programming GPIO in the node-red is made easy; the palette shows the permissible output and input pins to the programmers. A simple example to program a relay is given in the following section. Figure 6.17 shows the GPIO chart of the raspberry pi in palette section. The system configured pins are marked in red and black colors, the available pins are marked with green color.

3.3V Power - 1	2 - 5V Power
SDA1 - GPIO02 - 3	4 - 5V Power
SCL1 - GPIO03 - 5	6 - Ground
GPIO04 - 7	8 - GPIO14 - TxD
Ground - 9	10 - GPIO15 - RxD
GPIO17 - 11	12 - GPIO18
GPIO27 - 13	14 - Ground
GPIO22 - 15	16 - GPIO23
3.3V Power - 17	18 - GPIO24
MOSI - GPIO10 - 19	20 - Ground
MISO - GPIO09 - 21	22 - GPIO25
SCLK - GPIO11 - 23	24 - GPIO8 - CE0
Ground - 25	26 - GPIO7 - CE1
SD - 27	28 - SC
GPIO05 - 29	30 - Ground
GPIO06 - 31	32 - GPIO12
GPIO13 - 33	34 - Ground
GPIO19 - 35	36 - GPIO16
GPIO26 - 37	38 - GPIO20
Ground - 39	40 - GPIO21

Fig. 6.17 GPIO pin configuration for raspberry pi

Figure 6.18 elucidates the flow editor program of the 40th pin with the relay module. A Switch is configured and it is directly connected with the GPIO pin number 40.

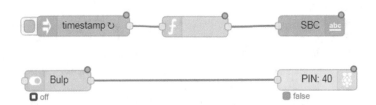

Fig. 6.18 GPIO pin programming with switch and time stamp in rpi

The user interface response of the above program is given in Fig. 6.19 below. Once the switch is toggled, the corresponding gpio pin 40 is triggered and the bulb is activated.

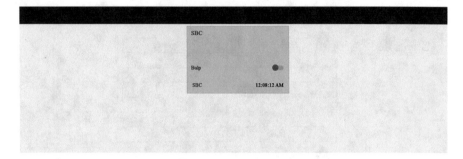

Fig. 6.19 Use Interface response of the bulb flow program

6.4   File Operations with Node-red

Figure 6.20 illustrates the file operation with the flow editor; the function block converts the mechanical clock data to readable format data. The time stamp is appended in the file and stored in the root folder of the raspberry pi. The same program can also be used to read data from the file and display the same in the user interface panel.

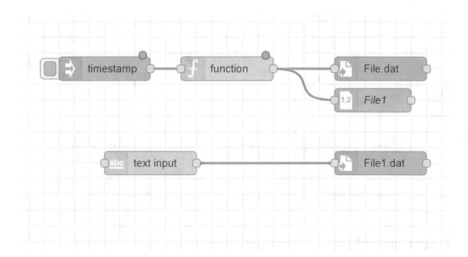

Fig. 6.20 Flow editor program for file storing operation

6.5   Terminal Window Command Operation with Node-red

The programmer can also pass terminal commands in the node-red, so that already developed program can also be triggered with the help of node-red flow editor. Figure 6.21 provides the switch control and exec block. The execution block to be filled with the command that is normally entered in the terminal block to run or activate a process say python filename.py is given. Figure 6.22 provides connection of switch control and exec block in flow editor.

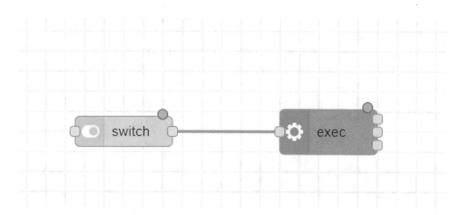

Fig. 6.21 Terminal command execution in node-red

Figure 6.22 illustrates the execution block command configuration.

Fig. 6.22 Exec block configuration in flow editor

6.6   Firebase Data Interface with Node-red

Firebase is the Google cloud free service; it can be directly accessed with the help of the node-red. Figure 6.23 elucidates the real-time database access with the Google firebase. Here text input block is directly connected with the firebase icon. The firebase module can be easily appended with the manage palette block set in window. The firebase address is filled in the free space and the authentication mechanism is to be anonymous for train projects. It can also be programmed with Gmail credentials. The read and write blocks in the firebase has to be set to true conditions. Figure 6.24 elucidates the configurations and necessary authentication details to be given in firebase block.

Fig. 6.23   Firebase flow editor with text input block set

Fig. 6.24   Firebase childpath configuration

After the process is done, the text input is fed with the ip:1880/ui window in browser and the text "123" is fed. The same out is immediately reflected in the firebase window, if both the sender and receiver have good internet connection. Figure 6.25 elucidates the firebase window elucidating the child data being updated.

Fig. 6.25 Firebase result window

6.7   Smtp, Email with the Raspberry Pi

The following section deals with the sending and reading an email with the node-red flow editor. This module helps to trigger a mail when some sensor senses events like fire or any other change in physical event. Figure 6.26 elucidates the details of the credentials and flow editor program in node-red window. Here the email block is connected with the pin 33.

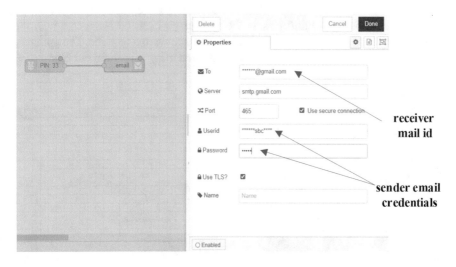

Fig. 6.26 Mail configuration in email palette

For a SBCs to send and receive mails via gmail, the less secure app access has to be enabled as given in Fig. 6.27. In some of the new boards, it is not necessary since it already provides https-based secure data exchange between the raspberry pi and Gmail server.

← Less secure app access

Some apps and devices use less secure sign-in technology, which makes
your account vulnerable. You can turn off access for these apps, which we
recommend, or turn it on if you want to use them despite the risks. Google
will automatically turn this setting OFF if it's not being used. Learn more

Allow less secure apps: OFF

Fig. 6.27 Figure Less secure application authentication in Google for successful node-red communication

6.8   Audio Out

The following section deals with text to speech conversion in the node-red. This module can be used in automated voice assistance in railway platforms, and other public announcements. Figure 6.28 elucidates simple text input block connected with the audio out, the audio jack in the raspberry pi can be connected with the speaker and the text given in the text block will be converted to voice. The option of voices like male and female can also be configured using this audio out block.

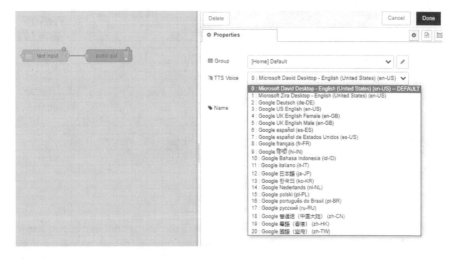

Fig. 6.28   Text to speech programming using node-red

6.9   PAGE Python GUI Builder

Python programming is mainly used for SBC prototyping and interfacing with real-time sensors. This section deals with building GUI and dashboard with drag-and-drop-based PAGE software [6, 7]. The page GUI builder is a Tool command language which creates python code after completion of design using drag and drop programming. The software can be downloaded from the link "https://sourceforge.net/projects/page/". Figure elucidates the basic gui window of the PAGE TCL software. More details on building the GUI is available in http://page.sourceforge.net/html/intro.html.

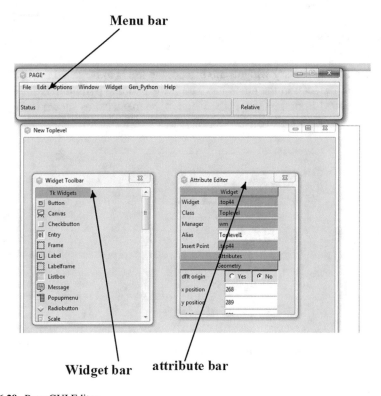

Fig. 6.29   Page GUI Editor

All the necessary widgets can be dragged from the widget window seen in the left side and can be placed in the required position in the template. The location, color, and other attributes of the widgets can be easily edited from the right hand attribute window. Figure 6.30 elucidates the widget being placed in the template using drag and drop methodology.

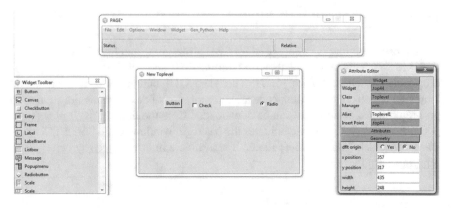

Fig. 6.30 Widget organized in PAGE software

Once all the necessary widgets are placed, the file is to be saved using the save option under file icon in the menu bar. The corresponding python files for the GUI can be generated from the Gen_Python icon in the menu bar as shown in Fig. 6.31.

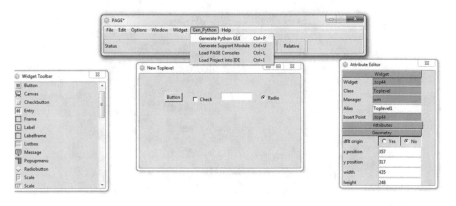

Fig. 6.31 Python code generation icon in Page menu bar

Both Python GUI and Support module has to be clicked, so that two python files will be generated in the page home folder. Figure 6.32 provides the python file generated after clicking the Generate Python GUI and Generate Support Module icons.

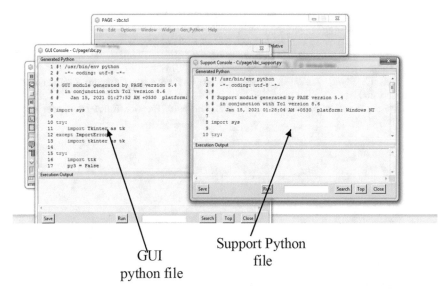

GUI
python file

Support Python
file

Fig. 6.32 Python codes built screen

The page software provides immediate easy gui building; these procedure can help and aid dashboard creation. This is a drag and drop tool and provides easy learning for intermediate programmers to build industry ready dashboards and can easily integrate with the existing platforms.

6.10 Conclusion

The explanation and programming of SBCs using node-red computers are clearly described in this chapter. Interface with sensors, cloud repositories, and actuators were discussed in detail. Python-based tkinter gui software is clearly described in this chapter.

References

1. M. A. Yoder, J. Kridner, *BeagleBone Cookbook: software and hardware problems and solutions* (2015). https://www.oreilly.com/library/view/beaglebone-cookbook/9781491915660/
2. A. Tamboli, *Build your own IoT platform: develop a fully flexible and scalable internet of things platform in 24 hours* (2019). https://www.oreilly.com/library/view/build-your-own/9781484244982/
3. S. Yamanoor, S. Yamanoor, U. Uthup, R. Pedley, S. Wood, J. Dharmaraj, *Raspberry Pi mechatronics projects HOTSHOT: enter the world of mechatronic systems with the rasp-*

berry Pi to design and build 12 amazing projects (2015). https://www.packtpub.com/product/raspberry-pi-mechatronics-projects-hotshot/9781849696227

4. D. Ibrahim, *Programming with node-RED*. S.l.: ELEKTOR (2020)
5. *Programming with tensorflow: solution for edge computing applications*. S.l.: SPRINGER (2020). https://www.springer.com/gp/book/9783030570767
6. C. Flynt, C. Flynt, *Tcl/Tk: a developer's guide*, 2nd edn. (Morgan Kaufmann, San Francisco, 2003)
7. B.B. Welch, *Practical programming in Tcl/Tk*, 4th edn. (Prentice Hall/PTR, Upper Saddle River, NJ, 2003)

Chapter 7
Cloud Interaction with SBCs

7.1 Introduction

Firebase is a free cloud platform where the data can be shared to real-time database. Most of the industries are still working with Google Spreadsheets and Google forms [1–8]. The sensors/actuators monitoring the industrial/physical events connected with the SBC can easily pump its data to the cloud service through this approach. Following section deals with the Google spreadsheet data sharing with the firebase. Figures 7.1, 7.2, 7.3, 7.4, 7.5, 7.6, 7.7, 7.8, 7.9, 7.10, and 7.11 depict the steps involved in Gsheets with firebase.

7.2 Sync Gsheets with Firebase

Step 1: Create your Firebase project

© The Author(s), under exclusive license to Springer Nature Switzerland AG 2021   141
G. R. Kanagachidambaresan, *Role of Single Board Computers (SBCs) in rapid IoT Prototyping*, Internet of Things, https://doi.org/10.1007/978-3-030-72957-8_7

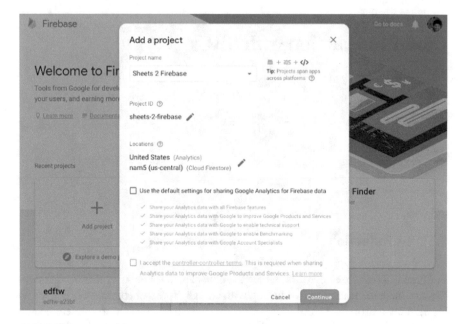

Fig. 7.1 Firebase creation in firebase.google.com

Step 2: Create your Realtime database
Navigate to Develop → Database and click the "Create database" button.

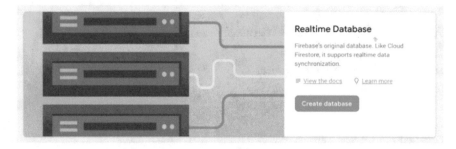

Fig. 7.2 Real-time firebase design

Make sure you change your read and write permissions to "true" and click publish.

```
1 ▾   {
2 ▾     /* Visit https://firebase.google.com/docs/database/security to learn more about security rules. */
3 ▾     "rules": {
4         ".read": true,
5         ".write": true
6       }
7     }
```

Fig. 7.3 Permission for realtime database editing

Editor's Note: While this is fine for initial development work, don't ever leave your security rules open like this in a production app. Make sure you lock them down, as the author notes later in the article.

Fig. 7.4 Security rules from Google

Copy down the database URL. It will be used in the below steps.

Fig. 7.5 Gsheets URL

The database url created is unique and can be called with the above link.
Step 3: Create your spreadsheet and populate it using this format

	id	firstName	lastName
1	1	Blake	Chapman
2	2	Ava	Sanderson
3	3	Boris	Robertson
4	4	Rachel	May
5	5	Anthony	Parsons
6	6	Olivia	Wallace
7	7	Justin	Carr
8	8	Joan	Baker
9	9	Nicholas	Ball
10	10	Gavin	Springer
11	11	Sally	Dowd
12	12	Adam	Morrison
13	13	Jasmine	Peake
14	14	Amy	Peters
15	15	Virginia	Henderson
16	16	Ella	Dyer
17	17	Ryan	Hardacre
18	18	Anna	Sutherland
19	19	Elizabeth	Manning
20	20	Colin	Vaughan
21	21	Edward	Hemmings
22	22	Alexander	North

Fig. 7.6 Google sheets data connectivity

The first row contains your keys. The first key should be set to "id" and each row should be labeled with the corresponding number, starting with "1." An easy way to set the id for each row in column A is to enter this formula "=COUNTA(B2:B2)" into cell A2 and then apply that to all rows. You can add as many rows or columns as you need.

Step 4: Create your Apps Script project
In the menu, go to Tools -> Script editor

Fig. 7.7 Appscript project
screen 1

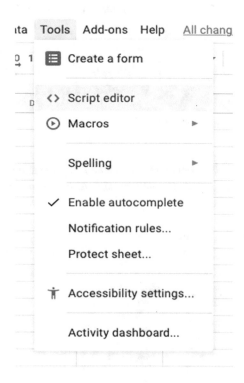

It will proceed to a code editor with the following file open: Code.gs. Replace the contents with this snippet. Search for this code at the top of the file:

```
function getEnvironment() {
  var environment = {
    spreadsheetID: '<REPLACE WITH YOUR SPREADSHEET ID>',
    firebaseUrl: '<REPLACE WITH YOUR REALTIME DB URL>'
  }
```

Fig. 7.8 Appscript project screen 2

Replace the "spreadsheetID" placeholder with users sheet address. The ID is the bolded part in the full spreadsheet URL (e.g. https://docs.google.com/spreadsheets/d/spreadsheetID/edit#gid=0)

Replace the "firebaseUrl" placeholder with your database URL from Step 2. Make sure to include the trailing slash (e.g., https://sheets-sample-test.firebaseio.com/); otherwise it will throw an error.

In your menu, go to View → Show manifest file, which will add a file called appsscript.json.

Fig. 7.9 Appscript project screen 3

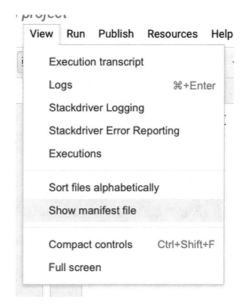

This will add an appsscript.json file to your project. Replace the contents with the following snippet.

Step 5: Start the sync

In the menu, go to Run → Run function → initialize. You will see a prompt to review and accept the permissions. This allows the App Script project to access the spreadsheet and upload data to Firebase. Click "Review Permissions" and then click "Allow."

Fig. 7.10 Appscript project screen 4

sheets to firebase wants to access your Google Account

This will allow sheets to firebase to:

●   See, edit, create, and delete your spreadsheets in Google   (i)
       Drive

       View and administer your Firebase Realtime Databases   (i)
       and their contents

       Connect to an external service   (i)

Make sure you trust sheets to firebase

You may be sharing sensitive info with this site or app. Learn about
how sheets to firebase will handle your data by reviewing its terms
of service and privacy policies. You can always see or remove
access in your **Google Account**.

Learn about the risks

Cancel   **Allow**

Fig. 7.11 Appscript authentication screen

The Firebase Realtime database has now been populated with the data from your spreadsheet! Any further edits will sync seamlessly and you can even share your spreadsheet with other people. Figure elucidates the data populated in the firebase screen (Fig. 7.12).

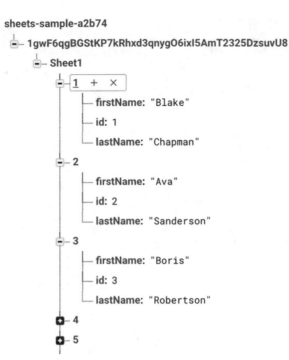

Fig. 7.12   Realtime firebase data screen

The next section deals with the google spreadsheet accessing with python through SBC. Jupyterlab notebook is installed in the SBC for demonstrating the below section.

7.2.1   Accessing Google Spreadsheet and Data Manipulation Using Python

Figure 7.13 provides the information on code snipped used to access the Google sheets. The pygsheets and numpy packages were used for this demonstration.

Fig. 7.13 Gspreadsheet module screen1

Follow the Google tutorials link for creating the client api and save the json as "client_secret.json." The screenshot is for the successful authentication of the client api.hotel. Figures 7.14, 7.15, and 7.16 provide screens to create and share the Gsheets with the third party using python.

7.3   Authentication

```
import os
import pygsheets
import numpy as np

# Authenticate API
gc = pygsheets.authorize('client_secret.json', service_account_file='client_secret.json')
print('API Authentication Successful')
```

Fig. 7.14 Authentication screen

7.3.1   Inserting Data and Sharing Gsheet

```
# Update a cell with value (just to let him know values is updated ;) )
wks.update_value('A1', "HELLO FOLKS")
my_nparray = np.random.randint(10, size=(5, 4))

# update the sheet with array
wks.update_values('A2', my_nparray.tolist())

# share the sheet with your friend
sh.share("myFriend@gmail.com")
```

Fig. 7.15 Code snippet screen 1 with share mail id

fx					
	A	B	C	D	E
1	HELLO FOLKS				
2	9	6	9	7	
3	9	8	6	8	
4	7	3	5	7	
5	4	7	7	8	
6	1	9	3	4	
7					

Fig. 7.16 Gsheet output window

7.3.2   *Inserting Random values*

Fig. 7.17 Interaction screen 1

The random code can be replaced with the connected sensor data or clock signal. Authenticating Firebase using Python is necessary; below figure provides the authenticating with Google screen (Figs. 7.18 and 7.19).

```python
import os
import pygsheets
import numpy as np

# Authenticate API
gc = pygsheets.authorize('client_secret.json', service_account_file='client_secret.json')
print('API Authentication Successful')
# Open spreadsheet and then worksheet
sh = gc.open('Test Sample')
wks = sh.sheet1

# Update a cell with value (just to let him know values is updated :) )
wks.update_value('A1', "HELLO FOLKS")
my_nparray = np.random.randint(10, size=(5, 4))

# update the sheet with array
wks.update_values('A2', my_nparray.tolist())

# share the sheet with your friend
# sh.share("myFriend@gmail.com")
```

Fig. 7.18 Random value insertion with python code

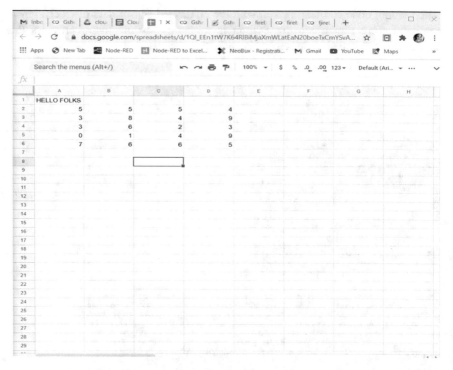

Fig. 7.18 (continued)

```
14 ∨ def signin_signup():
15      #global email, password
16      choice = input("Existing user?(Y/N): ")
17      email = input("Email: ")
18      password = input("Passw: ")
19
20 ∨  if choice == "n" or choice =='N':
21          '''New user'''
22          user = auth.create_user_with_email_and_password(email, password)
23          print(user)
24          print("User Account Created Successfully")
25          auth.send_email_verification(user['idToken'])
26          print("Please verify your email address")
27      '''Signin User'''
28      signin = auth.sign_in_with_email_and_password(email, password)
29      print("Signed Successfull")
30      c1 = input("Want to change password(y/n): ")
31 ∨  if c1 == "y":
32          changepassword()
33 ∨  else:
34          exit()
35 ∨ def changepassword():
36      print("Reset your password")
37      email = input("Email: ")
38      auth.send_password_reset_email(email)
39      print("Please check your email to change password")
40 ∨ if __name__ == "__main__":
41      global firebase, auth
42      firebase = pb.initialize_app(config)
43      print("Firebase succcessfully configured")
44      auth = firebase.auth()
45      print("Firebase authentication object successfully created")
46      signin_signup()
```

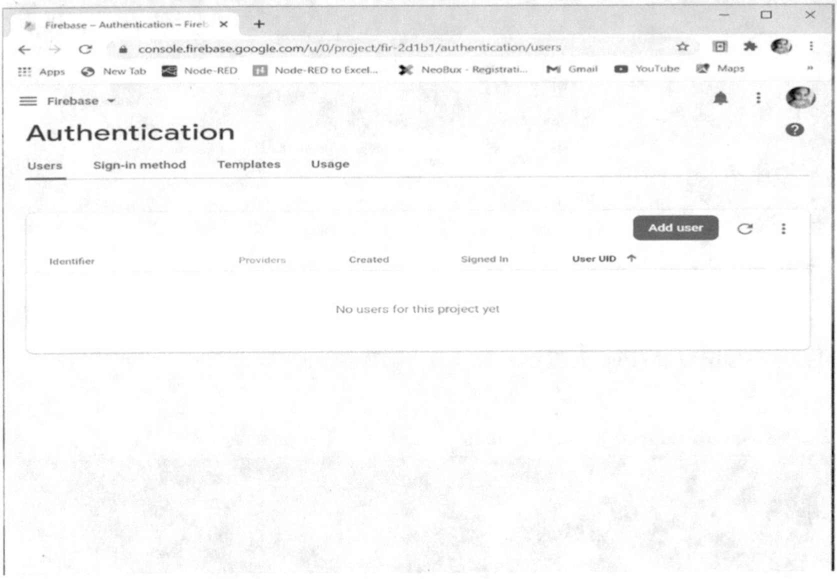

Fig. 7.19 Before Authenticating screen 1

Figure 7.20 provides the authentication mechanism for the above Python Code.

C:\Users\cpspu\AppData\Local\Programs\Python\Python36\python.exe

Firebase succccessfully configured
Firebase authentication object successfully created
Existing user?(Y/N): N
Email: iothacks@gmail.com
Passw: IoThack@123
{'kind': 'identitytoolkit#SignupNewUserResponse', 'idToken': 'eyJhbGciOiJSUzI1NiIsImtpZCI6ImUwOGI0NzM0YjYxNmE0MWFhZmE5Mm
N1ZTVjYzg3Yjc2MmRmNjRmYTIiLCJ0eXAiOiJKV1QifQ.eyJpc3MiOiJodHRwczovL3NlY3VyZXRva2VuLmdvb2dsZS5jb20vZmlyLLJKMWIxIiwiYXVkIjo
iZmlyLTJKMWIxIiwiYXV0aF90aW1lIjoxNjEwMDQyNDEwLCJ1c2VyX21kIjoiVHhobXZvZ1VuN1ZyQ2Q3YlVMbFdvNWZWRE1wMSIsInN1YiI6I1R4aG12b2d
VbjdWckNkN2JVTGxXbzVmVkRNcDEiLCJpYXQiOjE2MTAwMDI0MTAsImV4cCI6MTYxMDA0NjAxMCwiZW1haWwiOiJpb3RoYWNrc0BnbWFpbC5jb20iLCJlbWF
pbF92ZXJpZmllZCI6ZmFsc2UsImZpcmViYXNlIjp7ImlkZW50aXRpZXMiOnsiZW1haWwiOlsiaW90aGFja3MiY29tIl19LCJzaWduX2luX3Byb3Z
pZGVyIjoicGFzc3dvcmQifX0.kiN3JuYyIdYE6gHi5fvQq5mInXsXHdXkkujKI1vdnktyloj-GAyIhwjTYQ2YoI0blPrgHcIRRBQq7sQix64AvkFLJTb8OsL
E5GloP3QtlTKeIRPEIEKWocHGM_hic22xLzHCk2I8WGaQTbJM69G19iXbG1_8c5ta_XjeCGSQJmBFmjxmS4CEV0_aYiSWU8aqO-KnyMJdi6AAheXBo80oe3v
YJDv80_oZsIldzbaEoPxmRgQUtevDBnQNbI1BMZrcN0PWriQxinTAfL9TITOVoPnyp_JtK9Rd1GoT9bTkbkXtRilTQ212Kw_WmbUIAqNwKhmM3-cdvs4Op2
hXMKZeQ', 'email': 'iothacks@gmail.com', 'refreshToken': 'AG8BCndrdgsaXnDSnowie07oSg1gd65T7FEG-4tIQJKEfjtMW9ntJY7iDajN9y
99GYNzIVhjCerWMU5fuM35Zs-myIhkzThOX_2hdBj9B68J9BvNzgzRH_8R_xufcj8IzOUphuKaF1-_xmTlrPJjkSjEqtsBReHV3lpg9Vt17VjiPAy0JcbhvZ
eTPkktxzyuA8un1d-5n9ua', 'expiresIn': '3600', 'localId': 'TxhmvogUn7VrCd7bULlWo5fVDMp1'}
User Account Created Successfully
Please verify your email address
Signed Successfull

Process returned 0 (0x0)   execution time : 82.732 s
Press any key to continue . . .

Fig. 7.20 After authenticating Screen 2

In order to insert new users in the same project, follow the steps as given in Fig. 7.21.

Fig. 7.21 Authentication Screen 3

For existing user (Fig. 7.22):

C:\Users\cpspu\AppData\Local\Programs\Python\Python36\python.exe

Firebase succccessfully configured
Firebase authentication object successfully created
Existing user?(Y/N): Y
Email: iothacks@gmail.com
Passw: IoThack@123
Signed Successfull

Process returned 0 (0x0)   execution time : 18.537 s
Press any key to continue . . .

Fig. 7.22 Configuration success screen

The password authentication can be changed with code in order to remove third-party access after the completion of certain events. Figure 7.23 provides the steps involved in password change.

Fig. 7.23 Authentication access

The SBCs on downloading/taking images from rpi camera can upload the photos directly to firebase. Following section Figs. 7.24, 7.25, 7.26, 7.27, and 7.28 provide the steps involved in uploading files in firebase.

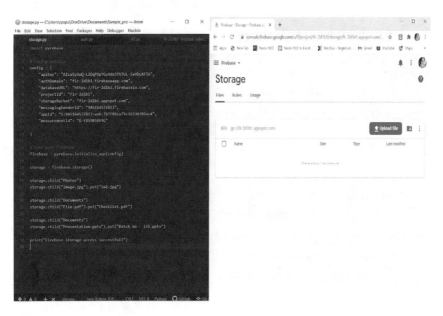

Fig. 7.24 File storage firebase screen 1

```python
import pyrebase

# Config setails
config = {
    "apiKey": "AIzaSyAwQ-L2DqPDpYGyAdn5fVJUL-SeVDyXF5k",
    "authDomain": "fir-2d1b1.firebaseapp.com",
    "databaseURL": "https://fir-2d1b1.firebaseio.com",
    "projectId": "fir-2d1b1",
    "storageBucket": "fir-2d1b1.appspot.com",
    "messagingSenderId": "945164572813",
    "appId": "1:945164572813:web:7b7f86ca79c16336f02ac4",
    "measurementId": "G-F8S9KSRE9G"

}

# Configure Firebase
firebase = pyrebase.initialize_app(config)

storage = firebase.storage()

storage.child("Photos")
storage.child("image.jpg").put("im2.jpg")

storage.child("Documents")
storage.child("File.pdf").put("Checklist.pdf")

storage.child("Documents")
storage.child("Presentation.pptx").put("Batch no - 135.pptx")

print("Firebase storage access successfull")

```

Fig. 7.24 (continued)

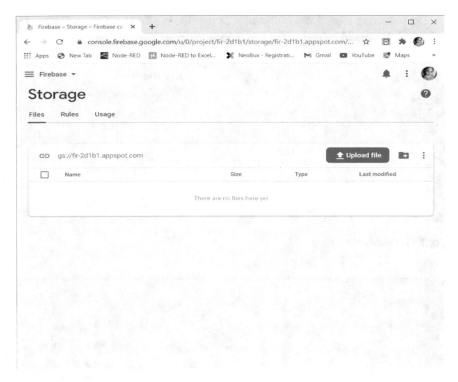

Fig. 7.24 (continued)

Following code snippet provides the information on image upload in firebase.

```
storage = firebase.storage()

storage.child("Photos")
storage.child("image.jpg").put("im2.jpg")

storage.child("Documents")
storage.child("File.pdf").put("Checklist.pdf")

storage.child("Documents")
storage.child("Presentation.pptx").put("Batch no - 135.pptx")

print("Firebase storage access successfull")
```

Fig. 7.25   File storage firebase screen 2

The following Figs. 7.26, 7.27, and 7.28 provide the information on folder creation and successful upload of the image from raspicam.

Fig. 7.26 Folder creation

Fig. 7.27 Image upload screen 1

Fig. 7.28 File upload screen 2

7.3.3 Insertion of Data in Realtime Database

The sensor data can be inserted into real-time database following the steps given below. The real-time database is free service and does not involve pricing to certain extent. The section starts with firebase.google.com, and following to real-time database section. The data can be uploaded and as well can be read from with the help of python in a SBC. The steps involved followed by the code snippet is elucidated in Figs. 7.29, 7.30, 7.31, 7.32, 7.33, 7.34, 7.35, and 7.36.

```
import pyrebase

# Config setails
config = {
    "apiKey": "AIzaSyAwQ-L2DqPDpYGyAdn5fVJUL-SeVDyXF5k",
    "authDomain": "fir-2d1b1.firebaseapp.com",
    "databaseURL": "https://fir-2d1b1.firebaseio.com",
    "projectId": "fir-2d1b1",
    "storageBucket": "fir-2d1b1.appspot.com",
    "messagingSenderId": "945164572813",
    "appId": "1:945164572813:web:7b7f86ca79c16336f02ac4",
    "measurementId": "G-F8S9KSRE9G"

}

# Configure Firebase
firebase = pyrebase.initialize_app(config)

db = firebase.database()

# Enter data into firebase

data = {"name": "Parwiz Forogh"}

db.child("abced").push(data)
print("Data added to real time database ")
```

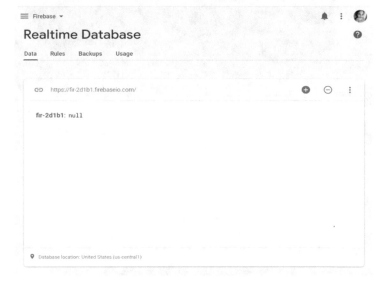

Fig. 7.29 Firebase real-time database screen 1

After uploading data

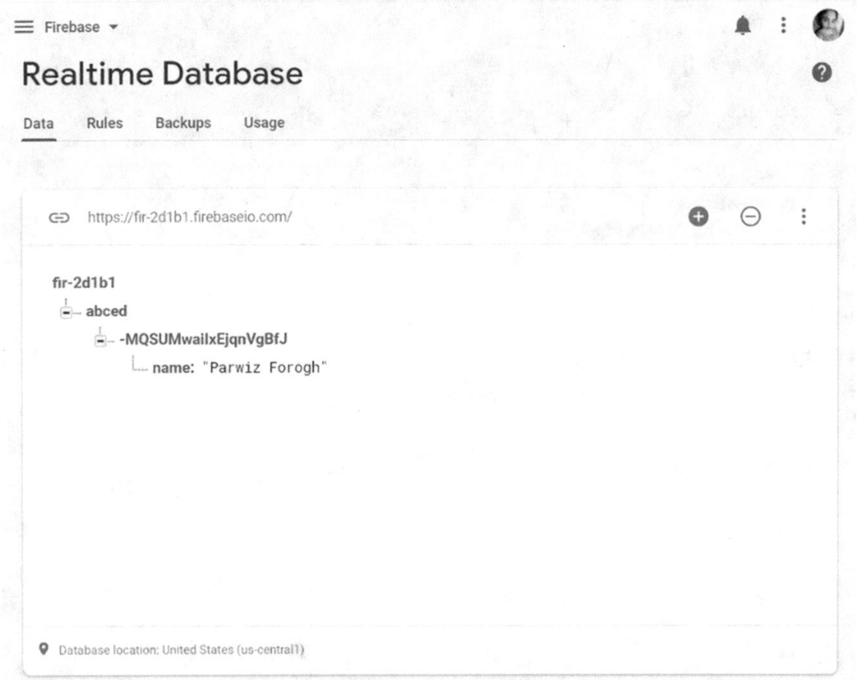

Fig. 7.30 Firebase realtime database screen 2

Fig. 7.31 Firebase realtime database code snippet

Fig. 7.32 Data with random parent file

The data can be extended with the existing parent file.

Fig. 7.33 Data appended with existing parent file using append command screen 1

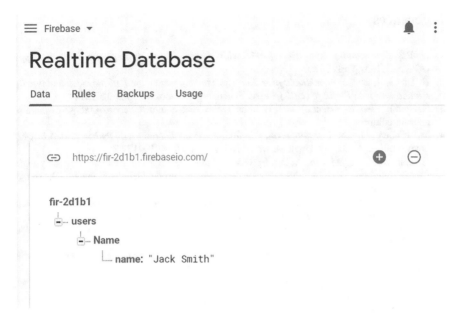

Fig. 7.34 Data appended with existing parent file using append command screen 2

```
# set
data = {"name": "Parwiz Forogh"}
db.child("users").child("Name").set(data)
# update
db.child("users").child("Name").update({"name": "Jack Smith"})
```

Fig. 7.35 Firebase Real-time database code snippet 1

```
# remove
db.child("users").child("Name").remove()
```

Fig. 7.36 Firebase Real-time database code snippet 2

Various possible steps for SBC to interact with cloud to send sensed/command data are elucidated with Python firebase package.

References

1. R. Payne, *Beginning app development with flutter: create cross-platform mobile apps* (APRESS, New York, 2020)
2. H. Yahiaoui, *Firebase cookbook: over 70 recipes to help you create real-time web and mobile applications with Firebase* (2017). https://dl.acm.org/doi/book/10.5555/3202514
3. N. Symth, *Firebase essentials* (2017). https://www.kobo.com/us/en/ebook/firebase-essentials-android-edition
4. M. Grinberg, *Flask web development: developing web applications with Python* (2018). https://www.oreilly.com/library/view/flask-web-development/9781491991725/
5. G. Ciaburro, V. Ayyadevara, A. Perrier, O.M.C. Safari, *hands-on machine learning on google cloud platform* (2018). https://www.packtpub.com/product/hands-on-machine-learning-on-google-cloud-platform/9781788393485
6. S.A. Kumar, *Mastering firebase for android development: build real-time, scalable, and cloud-enabled android apps with firebase* (Packt Publishing Ltd, Birmingham, 2018)
7. J. Shovic, *Python all-in-one for dummies*, 1st edn. (John Wiley and Sons, Indianapolis, IN, 2019)
8. L. Moroney, *The definitive guide to firebase: build android apps on Google's mobile platform*, 1st edn. (Apress: Imprint: Apress, Berkeley, CA, 2017)

Chapter 8
Introduction to KiCad Design for Breakout and Circuit Designs

8.1 Introduction

The pcb boards consist of many layers: (1) silk screen, (2) solder mask, (3) copper strate, and (4) substrate (FR4). The pcb boards can be designed with multiple layers, mainly double-layer and single-layer pcb boards are commonly used in IoT prototyping. Some of the commonly used terms in PCB_layout designs are solder mask, silkscreen, pads, layers, footprint, Jumpers, Via, Copper traces, etc. These notations are used to make user identification marks, circuits without short circuits and coloring options. PCB boards are mainly classified with through-hole design and layered pcb circuits; this chapter discusses on through-hole component design [1–3]. Figure 8.1 elucidates the kicad user interface boards with icons to proceed further.

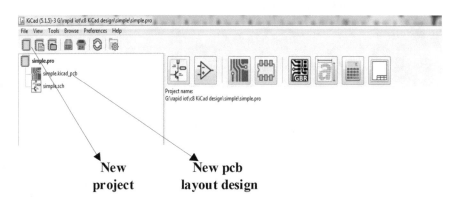

Fig. 8.1 User interface for KiCad

© The Author(s), under exclusive license to Springer Nature Switzerland AG 2021   165
G. R. Kanagachidambaresan, *Role of Single Board Computers (SBCs) in rapid IoT
Prototyping*, Internet of Things, https://doi.org/10.1007/978-3-030-72957-8_8

Figure 8.2 elucidates the icon options in the kicad design including important component designs, layer choice for routing circuits, wiring options, raster the pcb boards for better performance and Text option for user identification marks and comments.

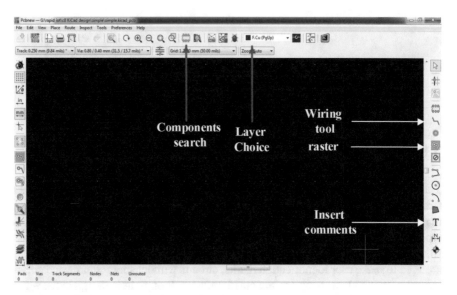

Fig. 8.2 Necessary navigation icons for Kicad

Figure 8.3 provides new foot print search in kicad software.

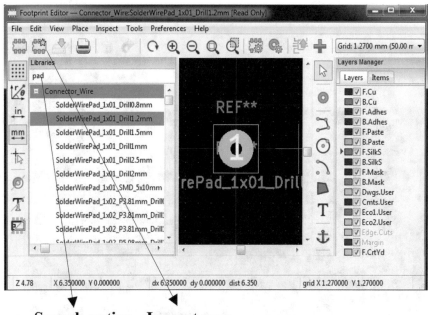

Search option   Import new
in library   foot print

Fig. 8.3 Foot print search in kicad

The circuit designed can be printed for real-time usage in IoT applications. The designed circuits can be printed in lab level with economic printers like LPKF printers. These options are temporary and can be done for very limited quantities. Figure 8.4 provides the LPKF printer loading and its components to print a PCB board. These printers are milling based and are limited to single- and double-sided pcb boards.

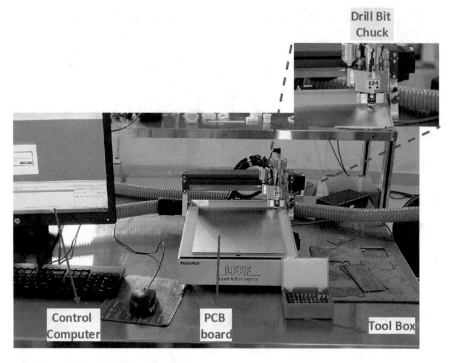

Fig. 8.4 LPKF printer for Lab scale PCB printer

The Gerber files generated from the kicad module is loaded in the LPKF software, and procedure initialization option is started. Figures 8.5 and 8.6 elucidate the gerber file loaded in the LPKF user window.

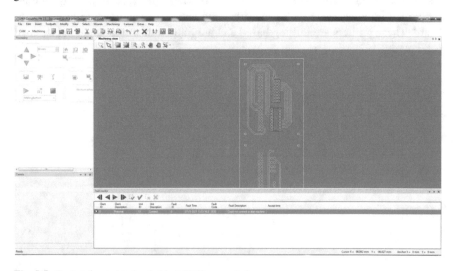

Fig. 8.5 Gerber front side loaded in LPKF user window

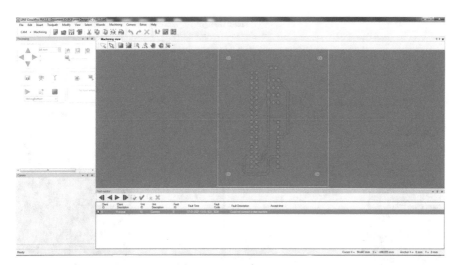

Fig. 8.6 Gerber Back side loaded in LPKF user window

The procedure is started and manual operator is in need to change the tools and tighten the chuck without disturbing the alignment of the pcb board mounted. First, the circuits are verified with the breadboard module and the same is designed in pcb design software like kicad, orcad, etc. Once the circuit is verified, the pcb is printed based on the quantity with lab level printers or with industry-ready pcb designers. Figure 8.7 elucidates the flow of PCB option from circuit design to industry-ready board making.

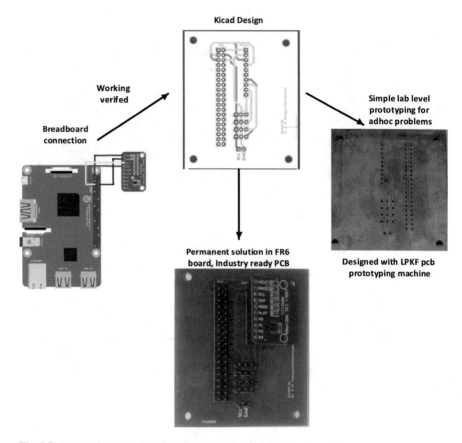

Fig. 8.7 Prototyping procedure from breadboard to industry-ready module

In this section, some of the mainly used breakout boards are designed; the reader can also do their own design following these examples. The i2c-based ads1115 breakout board front and back side is represented in Figure 8.8a and b. The size of the pcb board is maintained as small as possible to have economic boards.

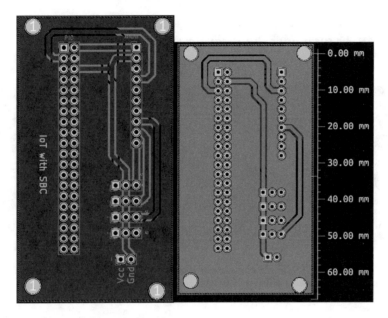

Fig. 8.8 (a) Front view. (b) Back view

8.2 ADS1115 Breakout Front and Back Kicad Circuit Design

Once the circuit is ready, the entire circuit can be viewed with the 3d viewer. Figure 8.9 represents the front and back 3d view of the breakout board. The same file can be transported to mechanical design software and assembly modules can be done (Figs. 8.10 and 8.11).

Fig. 8.9 3d view of the PCB boards

8.3   Color Sensor and HC-05 Device Integration for Colorimetric Sensing Breakout Board

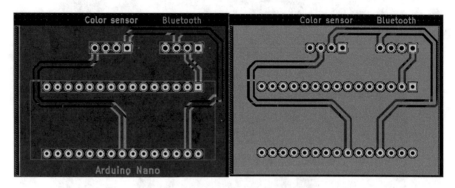

Fig. 8.10 Front view TCS34725 color sensor back view TCS34725 color sensor MCP3008 8 channel Raspberry pi breakout board, ki cad design

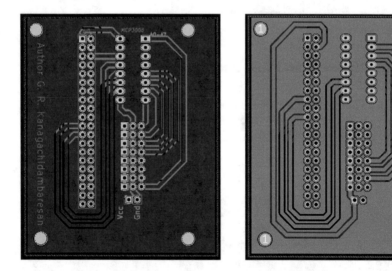

Fig. 8.11 MCP3008 front breakout design MCP3008 backside breakout design Xbee-based relay control unit for Industry 4.0

Industry relay operation boards with 5-channel and 6-channel custom-made design for 5A, 250 Volt machinery control and scheduling. Figure 8.12 elucidates the Kicad design of the proposed board.

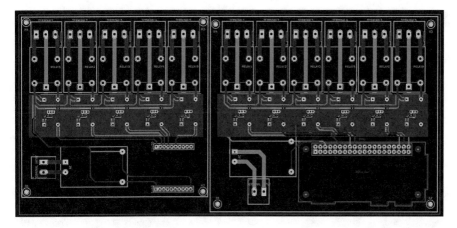

Fig. 8.12 PCB view of the xbee control board

Figure 8.13 provides the 3d view of the proposed industry relay control board.

Fig. 8.13   3d view of the proposed industry board

Figure 8.14 provides the fr4 pcb printed board for testing and monitoring purpose

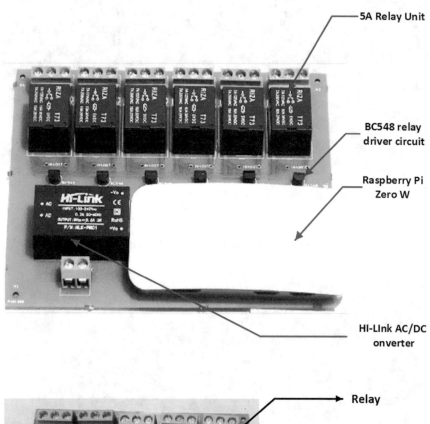

Fig. 8.14 Proposed Industry relay board printed in fr6 and assembled

References

1. S. Saponara, A. De Gloria (eds.), *Applications in Electronics Pervading Industry, Environment and Society: APPLEPIES 2018*, 1st edn. (Springer International Publishing: Imprint: Springer, Cham, 2019)
2. M. ParejaAparicio, *Diseño y desarrollo de circuitosimpresos con KiCad* (RC Libros, Madrid, 2010)
3. J. Pierre, T.C. Fabrizio, W. Stambaugh, *Kicad complete reference manual* (12th Media Services, Suwanee, 2018)

Chapter 9
Introduction to 3d Printing and Prototyping

9.1 Introduction

CATIA is the widely used tool to model 3d components for 3d printing. CATIA tool provides modeling of different designs from part design to generative sheet metal design. Among various such designs, ad hoc solutions for IoT and prototyping mainly needs part design and part assembly modules used in CATIA [1–5, p. 5]. This section deals with understanding the basic GUI components and modeling simple components for prototyping purpose. The chapter also deals with simple 3d printing concepts and also gives glimpse on existing 3d printing methods and its applications. Figure 9.1 provides the basic gui icons and tabs available in CATIA v5.

© The Author(s), under exclusive license to Springer Nature Switzerland AG 2021
G. R. Kanagachidambaresan, *Role of Single Board Computers (SBCs) in rapid IoT Prototyping*, Internet of Things, https://doi.org/10.1007/978-3-030-72957-8_9

9.2 Basic GUI Components of CATIA

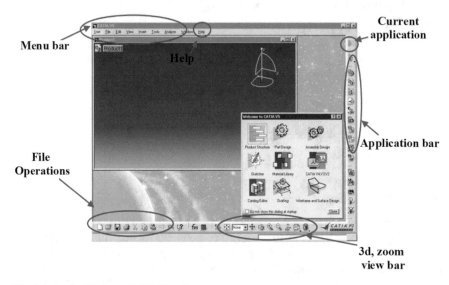

Fig. 9.1 Basic GUI icons in CATIA v5

It is very important to understand the GUI basics on CATIA for amicable modeling. Figure 9.2 elucidates starting a CATIA with mechanical part design model. This section deals about designing a casing for mounting water quality sensor as discussed in the previous chapter. Here the sensors, pH, dissolved oxygen, temperature, and turbidity sensors are considered.

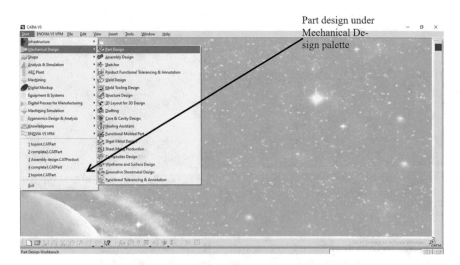

Fig. 9.2 Start screen with mechanical part design

Once the part design is chosen, the part what is to be designed has to be named. Figure 9.3 provides the window to provide user input name for the part file.

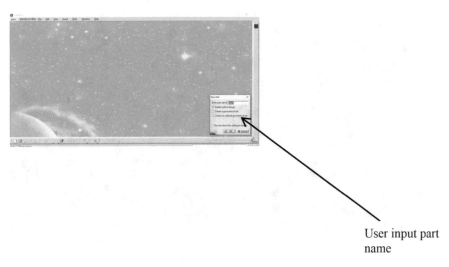

User input part name

Fig. 9.3 User input naming the part file

The screen proceeds with choice of plane in which the design operations will be carried. Figure 9.4 provides the option to choose the plane in which the base design will be carried out.

Plane selection palette

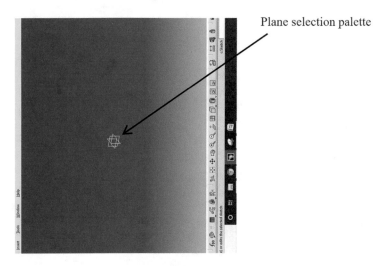

Fig. 9.4 Plane yz selection in CATIA window

Once the plane selection option is completed, a simple rectangle is drawn with necessary dimensions length and breadth as given in Figs. 9.5 and 9.6.

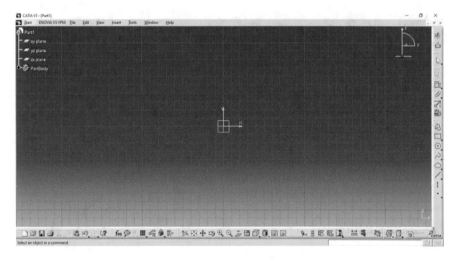

Fig. 9.5 2d yz plane

Figure 9.6 provides the rectangle placed with x, y user dimension in yz plane. The rectangle icon can be chosen as per the given figure.

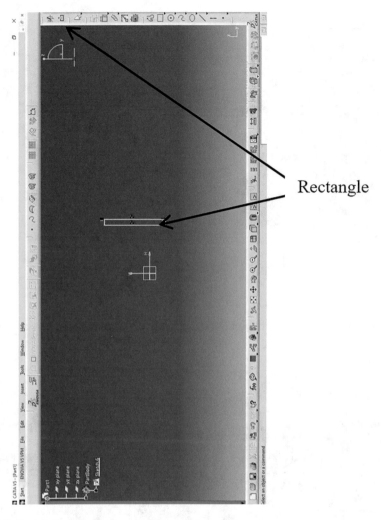

Rectangle

Fig. 9.6   Rectangle screen1

Once the rectangle is drawn with necessary dimensions, shafting is carried out for 360 to form a hollow cylinder-like structure. Figure 9.7 provides the details of shafting options and other necessary steps to be carried out.

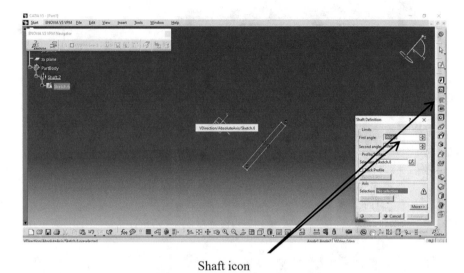

Shaft icon

Fig. 9.7   Shaft icon screen

After the choice of shafting is selected, the 360 view in xyz plane is seen as given
in Fig. 9.8.

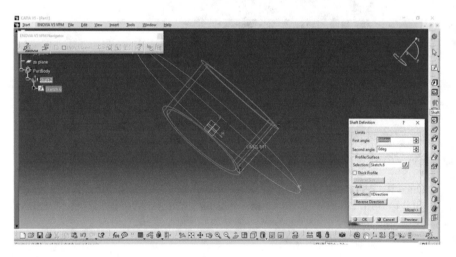

Fig. 9.8   Shaft selection screen

Confirming the shaft selection will yield a closed structure of the hollow cylinder
as given in Fig. 9.9.

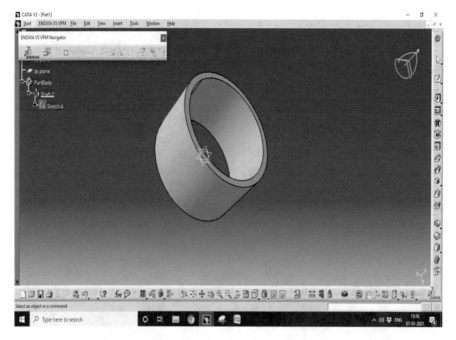

Fig. 9.9   Hollow cylinder screen

A padding is carried out to mount the sensors in the hollow cylinder as given in Fig. 9.10.

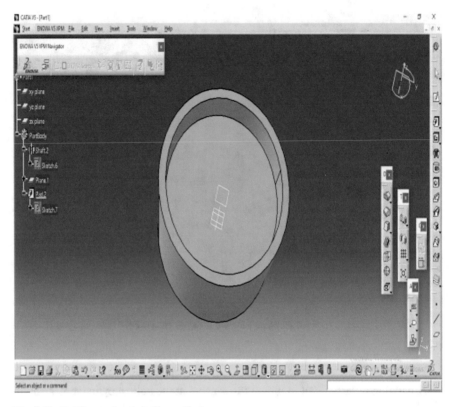

Fig. 9.10 Padding a plate in hollow cylinder

After padding process, necessary pocket is developed in the plate to mount pH, dissolved oxygen, temperature, and turbidity sensor. Figure 9.11 provides the steps involved in making a pocket in plate.

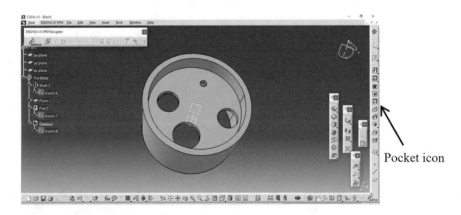

Fig. 9.11 Pocket to place Sensors

After completing the design, 3d view of the model can be seen by clicking icon as given in Fig. 9.12.

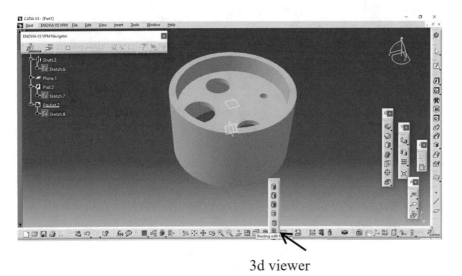

3d viewer

Fig. 9.12 Completed 3d model

Following section deals about a simple casing design for SBC, and part assembly procedure in CATIA.

9.3 Part Assembly Procedure for Prototyping

Figure 9.13 provides the basic 3d design as discussed earlier. In case of part assembly procedure, the 3d designs have to be developed in the part design and assembled in part assembly module separately. Hence, here, three files are developed, upper cover, lower chamber, and 4 screws for mounting procedure. Figures 9.13 and 9.14 show the lower part for mounting an arduino breakout board. The wall thickness is given as 7 mm with 6 mm diameter for screw mounting. The part is designed with repeated padding and pocket steps with user-defined width and breath.

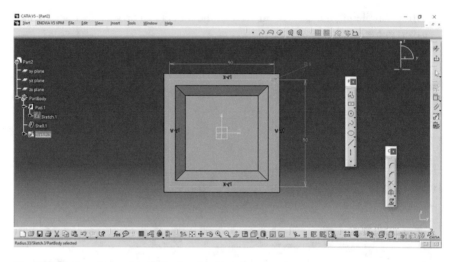

Fig. 9.13 Bottom design for arduino breakout board

Once the solid block is designed, mounting holes are placed in four edges of the block with 5 mm depth. Figure 9.14 provides the screen after pocket with circle icon.

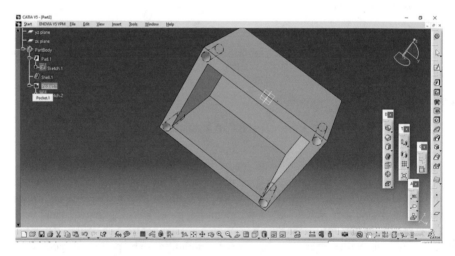

Fig. 9.14 Mounting screw hole screen

Figures 9.15, 9.16, and 9.17 provide the upper part cover designed with padding option and screw designed with the helix icon. The input pitch, diameter, etc. are mentioned in helix option window.

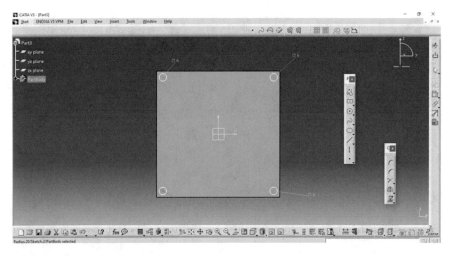

Fig. 9.15 Upper cover screen 1

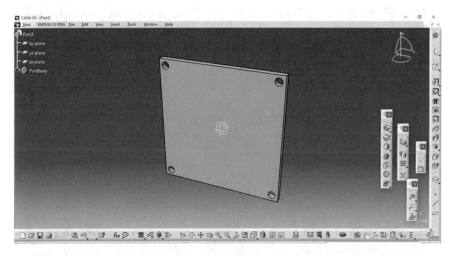

Fig. 9.16 Upper cover screen 2 with mounting holes

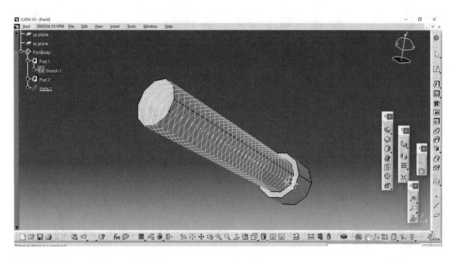

Fig. 9.17 Screw design with whirls

Once all the individual designs are completed, the files are saved with necessary names. A new part assembly files is initiated in mechanical design, with the import option all the part design files, lower, upper, screws are imported. Details are given in Figs. 9.18, 9.19, 9.20, and 9.21.

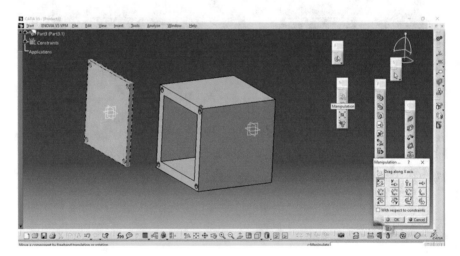

Fig. 9.18 Part assembly screen

The faces of the parts are chosen with control click; the planes to be merged are chosen with control click button and merged as given in Figs. 9.19, 9.22, and 9.23.

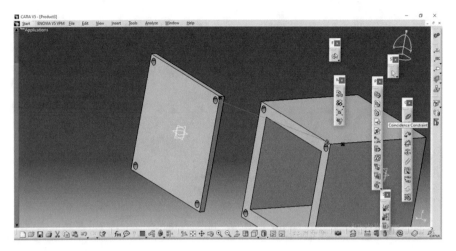

Fig. 9.19 Face merging screen

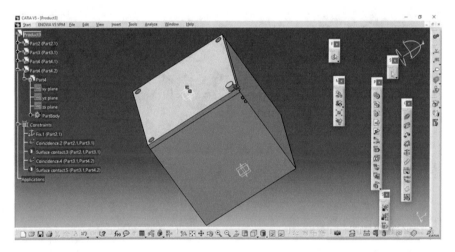

Fig. 9.20 After merging screen

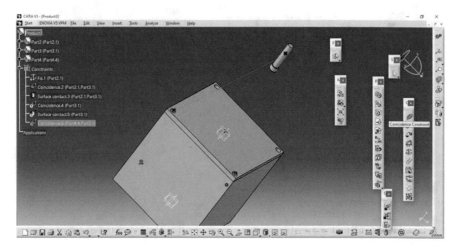

Fig. 9.21 Screw import screen 1

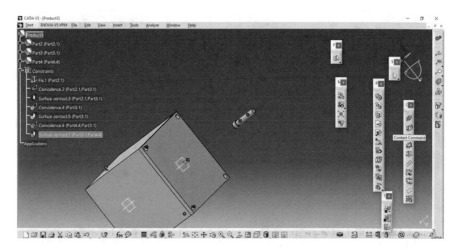

Fig. 9.22 Screw import screen 2

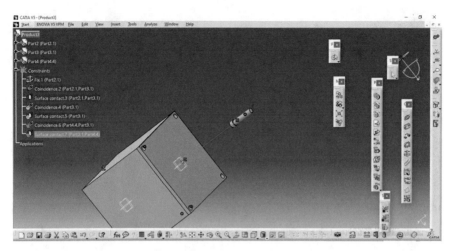

Fig. 9.23 Screw import screen 3

Once all the screws are properly faced with adjacent merging planes, assembling can be initiated. Figures 9.24 and 9.25 provide the blast view and assembled view of the design.

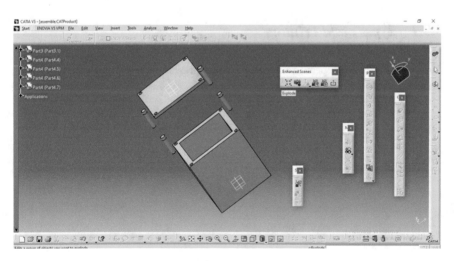

Fig. 9.24 Complete screw import screen

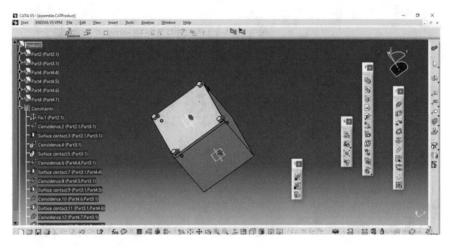

Fig. 9.25 Assembled block screen

Once the design is done and verified in CATIA, the same is exported as STL file for real-time printing in 3d printers. Following section provides the information on basics of 3d printing.

9.3.1 Types of 3d Printing

Type of 3d printing approaches mainly seen in present industry as well as academia includes [6–14]:

- Fused Deposition Model (FDM)

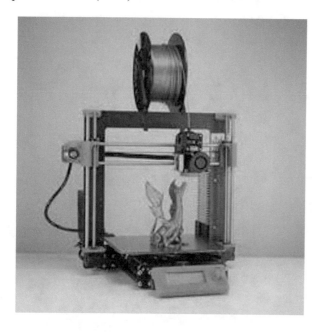

- Selective Laser Melting (SLM)

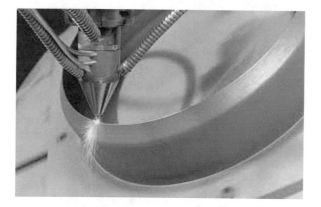

- Direct Laser Metal Sintering (DMLS)

- Selection Laser Sintering (SLS)

- Electronic beam melting (EBM)

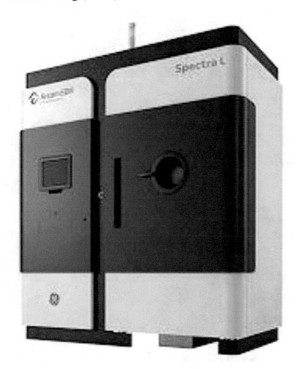

- Electronic Beam Additive Manufacturing (EBAM)

Among the mentioned 3d printing methods, FDM and SLS are commonly and widely used to satisfy economic immediate requirement. These printers can directly take the designed STL file and can initialize printing procedure. These printers play a major role satisfying major and daily needs in many areas. Figure 9.26 elucidates the dominant areas were 3d printing contributes daily needs.

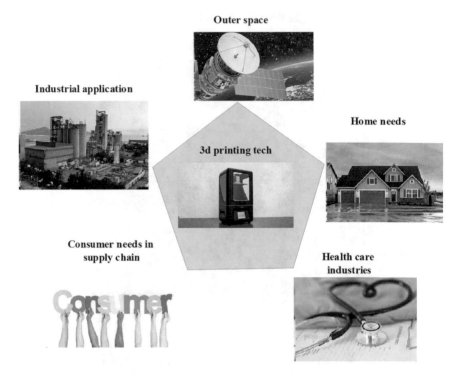

Fig. 9.26   3d Printing application domains

References

1. K. Plantenberg, *An introduction to CATIA V5: release 19; (a hands-on tutorial approach)* (Schroff Development Corp., Mission, Kan., 2009)
2. R. Cozzens, *CATIA V5 workbook: release V5-6R2013* (2013)
3. R. Cozzens, Schroff Development Corporation, *CATIA V5 Workbook: releases 19* (SDC Publications, Mission, Kan., 2009)
4. CADFolks, *CATIA V5-6R2014 for beginners* (Amazon, Marston Gate, 2014)
5. *CATIA V5-6R2018 for designers* (CADCIM TECHNOLOGIES, 2019). https://www.cadcim.com/catia-v5-6r2018-for-designers
6. A.K. France, *Make: 3D printing* (2014). https://www.oreilly.com/library/view/make-3d-printing/9781457183577/
7. G. Fisher, *Blender 3D Printing Essentials* (Packt Publishing, Birmingham, 2013). https://www.packtpub.com/product/blender-3d-printing-essentials/9781783284597
8. M. Rigsby, *A beginner's guide to 3D printing: 14 simple toy designs to get you started* (Chicago Review Press, Chicago, IL, 2014)
9. J.F. Kelly, *3D printing: build your own 3D printer and print your own 3D objects* (Que, Indianapolis, IN, 2014)

10. M. Ritland, *3D printing with SketchUp: real-world case studies to help you design models in SketchUp for 3D printing on anything ranging from the smallest desktop machines to the largest industrial 3D printers* (Packt Publishing, Birmingham, 2014)
11. J. Larson, *3D printing blueprints: design successful models for home 3D printing, using a Makerbot or other 3D printers* (Packt Publishing, Birmingham, 2013)
12. T. O'Neill, J. Williams, *3D printing* (Cherry Lake Publishing, Ann Arbor, MI, 2013)
13. J.M. Jordan, *3D printing* (The MIT Press, Cambridge, MA, 2018)
14. C. Coward, *3D printing* (2015)

Chapter 10
IoT Projects in Smart City Infrastructure

10.1 Introduction

Virtual reality-based training (VRBT) is an interactive and immersive teaching method that employs technology to provide virtual scenarios to simulate situations that might occur in actual settings. Virtual reality (VR) provides immersive user experience which makes them a cost-effective solution to employ for various training purposes [1–8]. This section uses virtual reality and machine learning in order to check the posturing and positioning of an amputated person with respect to normal healthy person. The people with amputated body parts come across a lot of difficulties due to the pain caused by the uneven amputation of legs or limbs. And sometimes this pain becomes unbearable for them. The frequency of prosthetic limb adjustment is identified using machine learning approach. The pose detection to detect the correct posture of the artificial legs or limbs of the amputated person is done through RPi cam and tensorflow package. Pose detection will be done in such a way that it will give accurate posture of the legs or limbs if legs or limbs are bend or damaged is detected. Figure 10.1 illustrates the posture angle for amputees estimated with gyroscope sensor mounted over his body.

© The Author(s), under exclusive license to Springer Nature Switzerland AG 2021
G. R. Kanagachidambaresan, *Role of Single Board Computers (SBCs) in rapid IoT Prototyping*, Internet of Things, https://doi.org/10.1007/978-3-030-72957-8_10

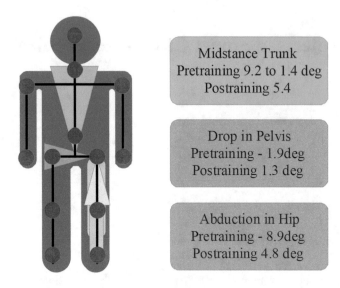

Fig. 10.1   Gait angle on post-training and pre-training

The methodologies mainly used are as follows:

Algorithm:

Step 1: Start

Step 2: Initiate Camera

Step 3: Detect pose with open3d pose detection package

Step 4: Compare the location with dataset and predict the error with the SVM

Step 5: Mark the points and map the stickman

Step 6: Provide corrective measure through mobile voice message and go to setp 3.

The key points identification and stickman realization is done with raspicam. The data of eyes, knees, and other necessary body faculties are trained with the tensorflow package. Figure 10.2 depicts the posture using tensorflow (Table 10.1).

Fig. 10.2 Achieved posture recognition with raspberry pi 4–8 Gb, raspicamera 5 MP

Table 10.1 The table elucidates the error in the posture by SBCs

S. No	Position	Error
1	P1	3
2	P2	2
3	P3	5
4	P4	3
5	P5	5
6	P6	2
7	P7	8
8	P8	2
9	P9	4
10	P10	4
11	P11	3
12	P12	5
13	P13	3
14	P14	6
15	P15	5
16	P16	5

Figure 10.3 provides the computational complexity of the above module with different SBCs.

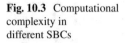

Fig. 10.3 Computational complexity in different SBCs

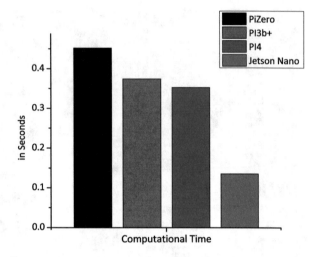

10.2   PLC Enabled Production Monitoring for Industry 4.0

The objective of this prototype is to work along with the existing PLC and record events to a database. Some of the industries are still operated with the old PLC device which does not support external interface such as RJ48, TCP, Ethernet, and Bluetooth facility. This prototype mainly converts the trigger signal and provides the same to dashboard.

The proposed module works along with PLC circuit and monitors the working of PLC and senses the events at 160 MHz frequency. The 24 V output of the PLC unit is divided to 5 V by series resistors and the noise signals are filtered using Zener diode. The Zener diode is fed inside the ADS1115 IC and Analog A0 pin of Node MCU. The node MCU unit is powered through charging adapter and programmed to feed the incoming data to the Google Spread sheet for further data processing and to monitor the production transparency. The equipment is also enabled with RS485 integrable device to directly communicate with the PLC unit. The data is received every 0.05 s, and the device is also configurable for different speed rates. The device is configurable to report regular and irregular testing mechanisms. The ADS1115 is a 4-channel 16 bit ADC converter; this is used to achieve high accuracy in sensing the event.

10.3   Introduction

The present testing and monitoring are being done through PLC systems [9–14]. The PLC systems are moreover not enabled with IoT and integrity with IoT devices is complicated [15–17]. The proposed programmable device is easily integrable with PLC device and can be configured for specific Google firebase and Google Sheet projects. The data is hopped to Google through Wi-Fi hotspot. The average timing to push data is less than a second. The device is presently configured with 4-channel ADC converter to monitor the data from sensors. Figure 10.4 illustrates the proposed programmable electronic circuit for monitoring data from machine.

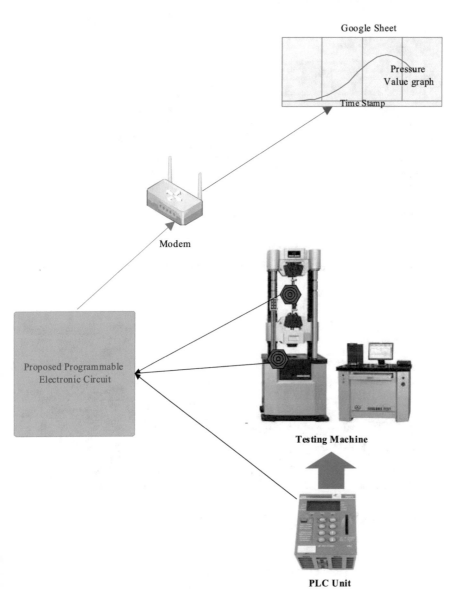

Fig. 10.4 Proposed Programmable Electronic Circuit

Figure 10.5 illustrates the Google sheet enabled with particular device. The Google sheet is integrated with particular Google sheet with Sheet id.

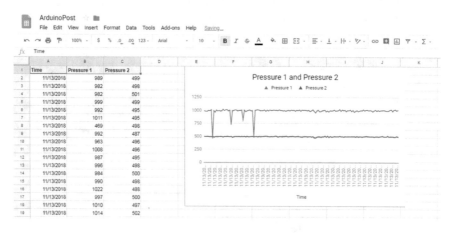

Fig. 10.5 Google Sheet displaying the current data

Figure 10.6 illustrates the Google sheet reporting all rig present production status.

Rig #	Product	Cycle	Target	% of Test Completion
Rig 1	Actutaor	146	1000000	0.01%
Rig 2	Actutaor		1000000	0.00%
Rig 3	Actutaor		1000000	0.00%
Rig 4	Actutaor		1000000	0.00%
Rig 5	Actutaor		1000000	0.00%

Fig. 10.6 Google Sheet displaying completed testing status

Figure 10.7 represent the proposed programmable electronic circuit.

Fig. 10.7 Front view of programmable electronic circuit used to monitor the values

The device is integrated with timer circuit that also reports the production time of the machine to the clients.

10.4   AI-Based Fruit Ripening Mechanism For Shipping and Logistics IoT Application

Around the world, different geographical locations and various temperate regions are favorable for the growth of varieties of fruits. In India, 97.35 Metric tons of fruits are produced, as per the release from agriculture ministry Oct 2019; the production is comparatively higher than the year 2018. India is the largest producer of fruits like banana, mango, guava, and lemon, and also one of the largest producers of orange, apple, and pineapple in the world. The price of the fruit is dependent on the taste, color, and shape, which is commonly referred as grade; good ripening results in best tasting fruits [18–23]. Each and every fruit has its own cultivation methods. The price of the fruit not only depends on the production cost, taste, grade or labor cost and profit, it also highly depends on the storage cost and shipping distance, considering these two highly influencing factors, AI-based fruit ripening mechanism for shipping and logistics IoT application device is designed. Ripening process increases the taste or sweetness of the fruit and the fruit becomes more palatable and softer when it ripens. Once the ripening process gets completed, it gets distributed from the storage to market or the food industries continue holding stocks under storage for several days analyzing the future fruit production and demand. The storage cycle of a fruit is a complex, time-consuming and cost-consuming process. Hence a module is designed to ripen the fruits within the container during the period of shipping, saving lot of time, space, and money, avoiding the onshore storage, and saving the environment by eliminating the extended refrigerated storage for ripening process even after shipping. Fruits can be directly transported to market or to distributers, understanding the world need and mass production.

10.5   Background

The fruit ripening is a cumulative process, where one fruit on ripening emits ethylene which triggers the other fruits to get ripened faster. The emission of ethylene is natural agent in ripening the fruits. The same is formed here, a pure nature ethylene gas is stored in the sprayer container and mounted on the containers carrying the fruits. The mounted sprayer device is controlled by IoT equipment with GPS sensor for perfect ripening of the fruit at particular location. This methodology mainly saves the time involved in storage and ripening, where the ripening is possibly done during the transportation.

10.6   Method

This GPS-based device can automatically spray the ripening gas during the period of transportation, thus integrating the process of transportation, storage, and ripening. Once the user input location matches the GPS serial data, spraying mechanism gets triggered, the predefined location plays a major role, the triggering location is set in a way that it was away from the destination. Hence it will start the ripening process within the container, few days before arriving the destination favorable for direct distribution and sales. Ethylene is a safe and accepted way of ripening fruits. A spraying mechanism is used to spray the gas from the spray can. It is mounted on the top of the ethylene gas spray can, once the location is detected the mechanism dispense the spray from the knob. When the container is some days away from the destination.

10.7   Detailed Description

10.7.1   Design of Controller

AI-based fruit ripening system is used to automatically ripen the fruits during the period of shipping or transportation. Ripening process is done using ethylene (C_2H_4) gas which is a globally accepted and safe method and makes the fruits more palatable to consume; ethylene gas is a naturally formed gas containing carbon and hydrogen atoms; this gas controls and matures the plants growth rapidly, increasing the storage lifetime of the fruit. The prototype is to integrate two operations, by which we can avoid the refrigerated storage—for some fruits like grapes, berries, and citrus fruits which must be refrigerated as the storage temperature is low, and temperate regions need high refrigerated storage, which releases large amount of greenhouse gases and harmful chlorofluorocarbon.

The module consists of a controller or a processing unit, sensing unit, and spraying mechanism as seen in Fig. 10.8. User feeds the particular input location to the processor; the location is nothing but the triggering location from where the ripening process must begin. The sensing module and the spraying mechanism is connected to the processor, which combinedly acts as automatic fruit ripening system.

10.8 Block Diagram

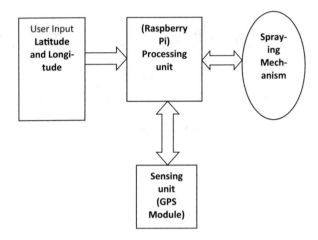

Fig. 10.8 Architecture of the system

Components used:

- Raspberry pi Zero w
- GPS module
- Stepper motor with driver
- Spraying mechanism (Rotational to linear motion mechanism)

10.9 Working

Automatic fruit ripening system uses GPS-based triggering. The fruit cultivation is mainly depending upon the transportation and shipping methods. According to the proposal to make the shipping more efficient, parallel processing is designed as shown in Fig. 10.9. The two methods are compared conventional process followed all these days and proposed automatic ripening method; the system is designed to integrate parallel processing of transportation and ripening for efficient transportation and shipping. As mobility is involved the best way to trigger spraying is to be analyzed as GPS triggering the flow of the process is shown in Fig. 10.8. The real-time simulation of user input, detection and activation of spraying mechanism when nearest location is sensed, the working of the system is serially monitored serially and shown in Figs. 10.9 and 10.10.

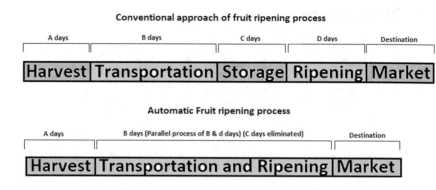

Fig. 10.9 Comparison of conventional and proposed model

Results: Thus, 1. Harvest > 2. transportation > 3. storage > 4. Ripening > 5. Distribution process is converted to 1. Harvest > 2. Transportation and ripening (parallel process) > 3. Distribution, which can result in more healthier fruits, as it is traveling directly to the market from fields, eliminating the period of storage and ripening after transportation. Table 10.2 provides functional components with its specification.

Table 10.2 The table illustrates the electronic components used and its function in the proposed patent's Hardware design

Components	Specification	Function
Raspberry pi	Zero w	Processing unit to which all the below mentioned components are connected through GPIO's. Programming language: Python OS: Raspbian OS
GPS Module	REES52 NEO-6M, GPS Positioning Module.	Sense the continuous current location using satellite coordinates.
Stepper motor with driver	28Ybj-48 5V DC motor With Uln2003 Driver	Drives the Spraying mechanism

Flowchart: Figure 10.10 illustrates the graphically represented flowchart of the proposed patent.

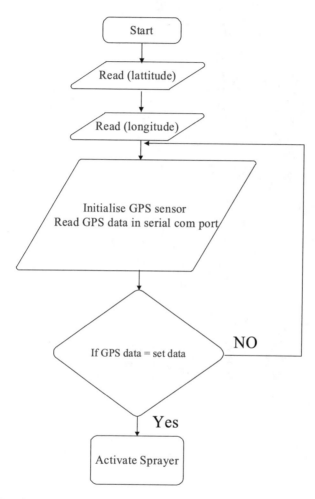

Fig. 10.10 Flowchart

Figure 10.11 illustrates the real-time hardware design, 3d printed component of the proposed prototype, and the components used in the module. Figure 10.12 provides the real-time proof of concept designed to check its accuracy. A screwing mechanism is 3d printed using FDM printer.

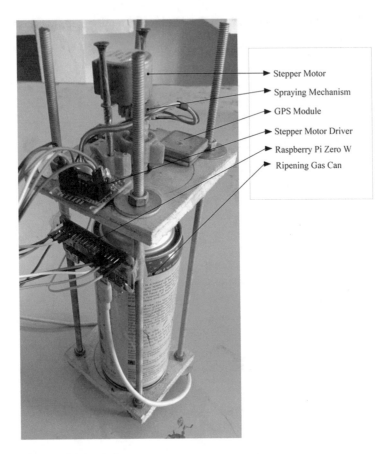

Stepper Motor
Spraying Mechanism
GPS Module
Stepper Motor Driver
Raspberry Pi Zero W
Ripening Gas Can

Fig. 10.11 Proposed proof of concept

Figure 10.12 illustrates the representation of Hardware components used in real time.

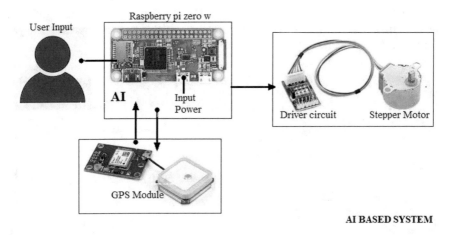

AI BASED SYSTEM

Fig. 10.12 Real-time components and architecture

Figure 10.13 depicts the real-time software simulation of the proposed patent, system prompts for triggering location and the user gives input.

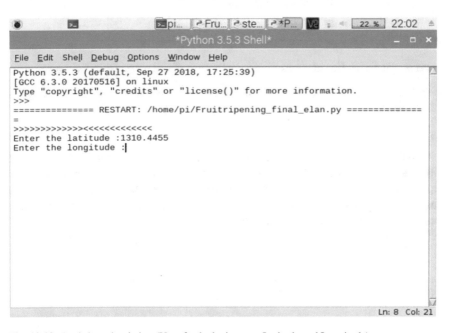

Fig. 10.13 Real-time simulation (User feeds the input as Latitude and Longitude)

Figure 10.14 depicts the real-time simulation of receiving continuous GPS data and the triggered spraying mechanism operation which sprays and releases the gas for programmed delay.

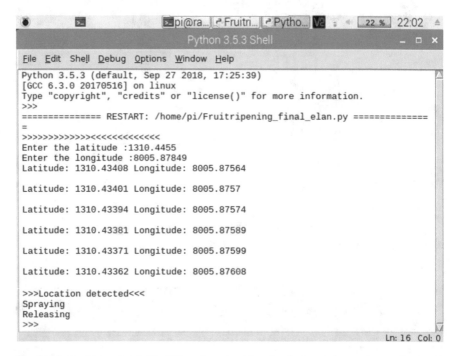

Fig. 10.14 Real-time simulation (When the nearest location is detected, spraying mechanism is activated)

Figure 10.15 illustrates the working of the 3D printed spraying mechanism. The illustration also gives a clear picture of real-time output when the system instructs to press the dispenser head. A rotational to linear converter 3D printed design is used in this proposed patent. The idle and the compressed state shows the linear movement of the pressing mechanism.

Fig. 10.15 Working of spraying mechanism

10.10   Conclusion

Industry 4.0 needs can be easily prototyped and deployed with the help of SBCs and available sensors. The 3d printing technology aids in providing immediate prototyping and ad hoc solutions to the industrial needs. This chapter provides the glimpse on raspberry pi interface with different sensors and actuators to satisfy smart city applications in the field of health care, industry 4.0, and supply chain.

References

1. D. Brulin, Y. Benezeth, E. Courtial, Posture recognition based on fuzzy logic for home monitoring of the elderly. IEEE Trans. Inf. Technol. Biomed. **16**(5), 974–982 (2012). https://doi.org/10.1109/TITB.2012.2208757
2. E.S. Sazonov, G. Fulk, J. Hill, Y. Schutz, R. Browning, Monitoring of posture allocations and activities by a shoe-based wearable sensor. IEEE Trans. Biomed. Eng. **58**(4), 983–990 (2011). https://doi.org/10.1109/TBME.2010.2046738
3. W. Ren, O. Ma, H. Ji, X. Liu, Human posture recognition using a hybrid of fuzzy logic and machine learning approaches. *IEEE Access* **8**, 135628–135639 (2020). https://doi.org/10.1109/ACCESS.2020.3011697
4. N. Khamsemanan, C. Nattee, N. Jianwattanapaisarn, Human identification from freestyle walks using posture-based gait feature. *IEEE Trans. Inf. Forensics Secur.* **13**(1), 119–128 (2018). https://doi.org/10.1109/TIFS.2017.2738611
5. Y. Yao, Y. Liu, Z. Liu, H. Chen, Human activity recognition with posture tendency descriptors on action snippets. *IEEE Trans. Big Data* **4**(4), 530–541 (2018). https://doi.org/10.1109/TBDATA.2018.2803838
6. Y. Chen, L. Yu, K. Ota, M. Dong, Hierarchical posture representation for robust action recognition. *IEEE Trans. Comput. Soc. Syst.* **6**(5), 1115–1125 (2019). https://doi.org/10.1109/TCSS.2019.2934639
7. J.-W. Wang, N.T. Le, C.-C. Wang, J.-S. Lee, Hand posture recognition using a three-dimensional light field camera. *IEEE Sens. J.* **16**(11), 4389–4396 (2016). https://doi.org/10.1109/JSEN.2016.2546556
8. A. Kleinsmith, N. Bianchi-Berthouze, A. Steed, Automatic recognition of non-acted affective postures. *IEEE Trans. Syst. Man Cybern. Part B Cybern.* **41**(4), 1027–1038 (2011). https://doi.org/10.1109/TSMCB.2010.2103557
9. M. Compare, P. Baraldi, E. Zio, Challenges to IoT-enabled predictive maintenance for industry 4.0. *IEEE Internet Things J.* **7**(5), 4585–4597 (2020). https://doi.org/10.1109/JIOT.2019.2957029
10. M. Aazam, S. Zeadally, K.A. Harras, Deploying fog computing in industrial internet of things and industry 4.0. *IEEE Trans. Ind. Inform.* **14**(10), 4674–4682 (2018). https://doi.org/10.1109/TII.2018.2855198
11. C. Garrido-Hidalgo, D. Hortelano, L. Roda-Sanchez, T. Olivares, M.C. Ruiz, V. Lopez, IoT heterogeneous mesh network deployment for human-in-the-loop challenges towards a social and sustainable industry 4.0. *IEEE Access* **6**, 28417–28437 (2018). https://doi.org/10.1109/ACCESS.2018.2836677
12. L.B. Furstenau et al., Link between sustainability and industry 4.0: trends, challenges and new perspectives. *IEEE Access* **8**, 140079–140096 (2020). https://doi.org/10.1109/ACCESS.2020.3012812
13. B. Aziz, Modeling and analyzing an industry 4.0 communication protocol. *IEEE Internet Things J.* **7**(10), 10120–10127 (2020). https://doi.org/10.1109/JIOT.2020.2999325

14. C. Marnewick, A.L. Marnewick, The demands of industry 4.0 on project teams. *IEEE Trans. Eng. Manag.*, 1–9 (2019). https://doi.org/10.1109/TEM.2019.2899350
15. S. Rani, R. Maheswar, G.R. Kanagachidambaresan, P. Jayarajan, *Integration of WSN and IoT for smart cities* (Springer, Cham, 2020)
16. G.R. Kanagachidambaresan, R. Maheswar, V. Manikandan, K. Ramakrishnan, *Internet of Things in smart technologies for sustainable urban development* (Springer, Cham, 2020)
17. *Advanced deep learning for engineers and scientists: a practical.* S.l.: SPRINGER NATURE (2021). https://www.springer.com/gp/book/9783030665180
18. W. Yang, Y. Xi, N. Yamauchi, Y. Miyazaki, N. Baba, H. Ikeda, "A Remote Wireless Networked Sensing System for Monitoring Stress of Fruits during Transportation," in *2006 SICE-ICASE International Joint Conference* (Busan Exhibition & Convention Center-BEXCO, Busan, Korea, 2006) pp. 3577–3580. https://doi.org/10.1109/SICE.2006.314728
19. G. Elavarasi, G. Murugaboopathi, S. Kathirvel, "Fresh Fruit Supply Chain Sensing and Transaction Using IoT," in *2019 IEEE International Conference on Intelligent Techniques in Control, Optimization and Signal Processing (INCOS)*, Tamil Nadu, India (Apr. 2019) pp. 1–4. https://doi.org/10.1109/INCOS45849.2019.8951326
20. Jianhua Yang and Jinjun Liu, "Management model investigation of fruit and vegetable supply chain centered on Third-Party Logistics Enterprise," in *2010 International Conference on Future Information Technology and Management Engineering*, Changzhou, China (Oct. 2010) pp. 108–111. https://doi.org/10.1109/FITME.2010.5654811
21. C. Zhu, Q. Wang, "Technical Measures of Fruit and Vegetable Transportation," in *2010 International Conference on Logistics Engineering and Intelligent Transportation Systems*, Wuhan, China (Nov. 2010) pp. 1–3. https://doi.org/10.1109/LEITS.2010.5664997
22. Y. Hongli, W. Yongming, "Transportation expenses minimal modelling with application to fresh food supply chain," in *2017 13th IEEE International Conference on Electronic Measurement & Instruments (ICEMI)*, Yangzhou, China (Oct. 2017) pp. 263–267. https://doi.org/10.1109/ICEMI.2017.8265786
23. Y. Kai, Z. Junmei, L. Wenbin, Y. Liu, G. Lin, X. Huixia, "Weighing System of Fruit-Transportation Gyrocar Based on ARM," in *2011 Third International Conference on Measuring Technology and Mechatronics Automation*, Shanghai, China (Jan. 2011) pp. 1146–1149. https://doi.org/10.1109/ICMTMA.2011.253

Chapter 11
Industry 4.0 for Smart Factories

11.1 Introduction

This chapter mainly focuses on several real-time projects that serve industrial needs. Monitoring of water quality parameters is mandatory in many industries to improve the production quality and quality. Several challenges rise on improving the quality of monitoring and maintaining accuracy in monitored data [1–3]. Industries are looking towards maintenance-free systems to satisfy their long-term needs [4–9].

11.2 Continuous Water Quality Monitoring with Self-healing IoT

11.2.1 Working Principle

The inlet water from water resources are flown through high-pressure electromagnetic pressure meters. The high-pressure electromagnetic pressure meters are equipped with rs485 protocol. A provision is made to collect the incoming water to the temporary small storage tank were the water sample is collected. The temporary storage tank is equipped with pH, temperature, turbidity, and dissolved oxygen sensors. Table 11.1 illustrates the sensor saturation time for the container (Volume 1531.311 cm^3).

© The Author(s), under exclusive license to Springer Nature Switzerland AG 2021   217
G. R. Kanagachidambaresan, *Role of Single Board Computers (SBCs) in rapid IoT Prototyping*, Internet of Things, https://doi.org/10.1007/978-3-030-72957-8_11

Table 11.1 Saturation time
of sensors

S. No	Sensor type	Saturation time (s)
1.	pH	26
2.	Temperature	25
3.	Dissolved oxygen	30
4.	Turbidity	65

Figure 11.1 illustrates the architecture of the proposed work where the FRDM boards NUCLEO F031K6 are installed in electromagnetic flow meters. The processors are connected with SPI communication with central CPU-NUCLEO FRDM-K64-ARM processor. The data is sent to dumb terminal like raspberry pi or other system for monitoring purpose. The quality of data is collected for every 2 min. The system is auto-cleaned for every 8 h.

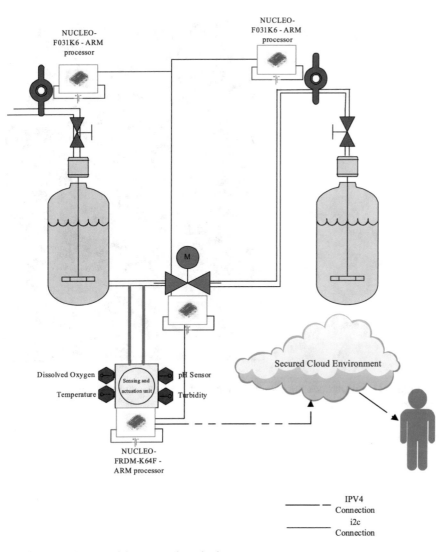

Fig. 11.1 Architecture of the proposed monitoring system

The entire mechanism is tested in lab condition with different water hardness. The sensors are auto-cleaned using acetone solution and the same is neutralized using the freshwater jet through submersible pump. The submersible small pumps are operated through relay operation with 9 v Input supply. The relay module is controlled by NUCLEO FRDM-K64-ARMF. Figure 11.2 illustrates the Proof of Concept developed for monitoring the system.

Fig. 11.2 Real-time
implementation of the
proposed work

Figure 11.3 illustrates the response graphs of the water and saturation time of the
sensors.

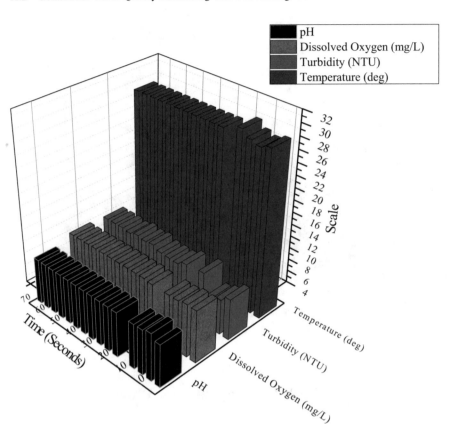

Fig. 11.3 Sensor response graphs

The round trip time taken to store the data in firebase cloud is given in Fig. 11.4.

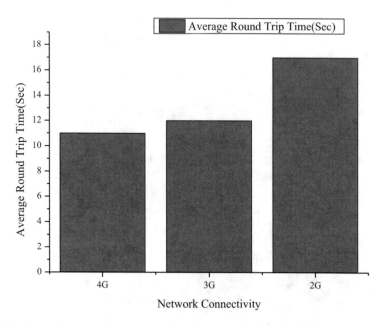

Fig. 11.4 Network response time

11.3   IoT Enabled Self-Cleaning Buoy for Water Quality Monitoring

The quality of water has an impact on the living beings. Water quality testing is an important part of environmental monitoring. Water quality refers to the chemical, physical, and biological characteristics of water. It is a measure of the condition of water relative to the requirements of one or more biotic species and or to any human need or purpose. In this section, the main parameters that define water quality are monitored and observed. To monitor the parameters, different sensors like pH and temperature are used. All the measured parameters are compared with the threshold value that defines the purity. Once the parameters are measured, they are sent to authority in the form of alert messages. Polluted water can cause devastating effects on the public health. It may even cause disease outbreaks. Around 4 million people die every year from water-related diseases.

Aquaculture has been a fast-growing industry because of significant increase in demand for fish and seafood. In aquaculture, the yields (shrimp, fish, etc.) depend on the water characteristics of the aquaculture pond. For maximizing fish yields, the parameters should be kept at certain optimal levels in water. The parameters can vary a lot during the period of a day and can rapidly change depending on the external environmental conditions. Hence it is necessary to monitor these parameters. India is leading in fish production and aquaculture organism exports. The feed conversion ratio and production of the fish is mainly determined by the water parameters like pH, DO, TDS, and temperature. Continuous monitoring is a challenging issue in coagulated environment; cloudy water particles make it even hard. Figure 11.5 elucidates the proposed water buoy design for continuous monitoring of pond water within minimal self-maintenance.

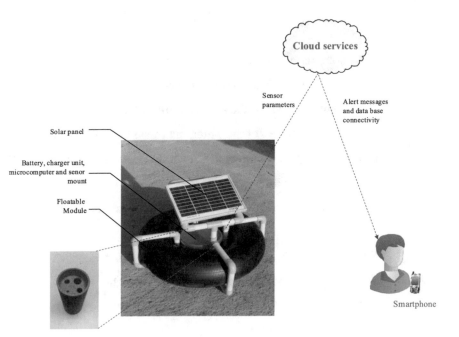

Fig. 11.5 Architecture of the proposed system

Sensors collect the data from the water and send it to the raspberry pi. Machine learning algorithm in Raspberry pi then analyzes the data from the sensors and send it to the Analog to Digital (A-D) converter and converts the encoded data into digital form. The collected data is received through the wi-fi module and stored in the cloud. Whole system is powered by the solar panel and it is connected to a rechargeable battery which supplies power to all the equipment within the system. This proposed module is a standalone unit which does not need any short-term maintenance.

This section has a floating module with water quality sensors mounted on with solar panel and Machine learning system. The system will identify water quality parameters every 45 s and reports the same to cloud through IoT module. A Machine learning algorithm is developed to identify the correctness of the sensor data. Android application is also developed for aqua farmers to get acknowledgement and notification on water quality parameters via smart phones.

A freshwater can is mounted in the buoy and is connected with small dc pump to pump the freshwater to the sensor mounting module. The sensor mounting module is designed with hollow tube with swirls and holes to create a jet of water to flush the coagulated dust items and other foreign materials inside the sensor mounting module.

Algorithm 1: Activation Algorithm

> **Start the process**
>
> If it is day time, power the module directly from solar and charge the battery through charger circuit
>
> If it is night time, power the module from battery back up
>
> Sense the parameters of the water and convert the analog data with the A/D converter, a pcb circuit board design with adc1115 and mcp3008 is developed.
>
> Run Machine learning algorithm to find the trust of the parameters.
>
> If the trust is high, share it to cloud through wifi connectivity or share it to remote location through wireless radio
>
> If the trust is low, flush the sensor mounting module and re-check the sensor content
>
> Store the data in local file
>
> Repeat the process from step one
>
> **End**

Figure 11.6 provides the steps involved in the water buoy design and monitoring.

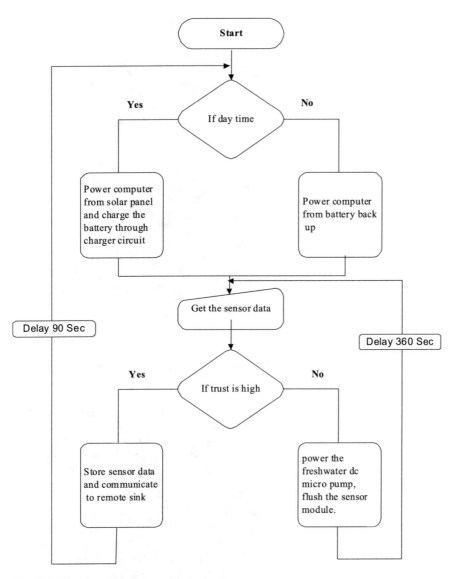

Fig. 11.6 Flowchart of the process involved

Figure 11.7 illustrates the 3D module design from different angles; the 3D design is generated using Solid works, a powerful designing software. The module has a height of 25 cm and width of 12 cm with an internal swirl design used for fetching and sampling the data. This design is printed using ULTIMAKER 3D printer simulated and positioned using ultimaker official software. The material used for printing this module is ABS (Acrylonitrile Butadiene Styrene).

Fig. 11.7 3D printed
proposed sensor mounting
module

Figure 11.8 provides the android application designed for the proposed work.

Fig. 11.8 Android User
Interface

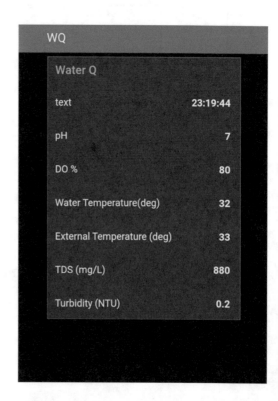

11.4 Conclusion

Role of SBCs on industrial monitoring to serve ad hoc immediate requirements was discussed in detail in this chapter. The end-to-end steps involved in prototyping were clearly depicted in this chapter. The chapter gives the overall picture on programming, circuit design, 3d printing, and android application to provide end-end solution for remote monitoring.

References

1. S. Rani, R. Maheswar, G.R. Kanagachidambaresan, P. Jayarajan, *Integration of WSN and IoT for smart cities* (Springer, Cham, 2020)
2. G.R. Kanagachidambaresan, R. Maheswar, V. Manikandan, K. Ramakrishnan, *Internet of Things in smart technologies for sustainable urban development* (Springer, Cham, 2020)
3. *Advanced deep learning for engineers and scientists: a practical.* S.l.: Springer Nature (2021)
4. A. Saad, A.E.H. Benyamina, A. Gamatie, Water management in agriculture: a survey on current challenges and technological solutions. IEEE Access 8, 38082–38097 (2020). https://doi.org/10.1109/ACCESS.2020.2974977
5. D.V. Cruz, M.R.G. de Oliveira, M.C. Filho, D.V. da Cruz, Monitoring pH with quality control based on Geostatistics Methodology. IEEE Lat. Am. Trans. 14(12), 4787–4791 (2016). https://doi.org/10.1109/TLA.2016.7817012
6. S.O. Olatinwo, T.-H. Joubert, Energy efficient solutions in wireless sensor systems for water quality monitoring: a review. IEEE Sens. J. 19(5), 1596–1625 (2019). https://doi.org/10.1109/JSEN.2018.2882424
7. S.O. Olatinwo, T.-H. Joubert, Enabling communication networks for water quality monitoring applications: a survey. IEEE Access 7, 100332–100362 (2019). https://doi.org/10.1109/ACCESS.2019.2904945
8. N.A. Cloete, R. Malekian, L. Nair, Design of smart sensors for real-time water quality monitoring. IEEE Access 4, 3975–3990 (2016). https://doi.org/10.1109/ACCESS.2016.2592958
9. D. Madeo, A. Pozzebon, C. Mocenni, D. Bertoni, A low-cost unmanned surface vehicle for pervasive water quality monitoring. IEEE Trans. Instrum. Meas. 69(4), 1433–1444 (2020). https://doi.org/10.1109/TIM.2019.2963515

Index

© The Author(s), under exclusive license to Springer Nature Switzerland AG 2021
G. R. Kanagachidambaresan, *Role of Single Board Computers (SBCs) in rapid IoT Prototyping*, Internet of Things, https://doi.org/10.1007/978-3-030-72957-8

Printed in the United States
by Baker & Taylor Publisher Services